Life's Other Secret

Life's Other Secret

THE NEW MATHEMATICS OF THE LIVING WORLD

Ian Stewart

ALLEN LANE
THE PENGUIN PRESS

ALLEN LANE
THE PENGUIN PRESS

Published by the Penguin Group
Penguin Books Ltd, 27 Wrights Lane, London W8 5TZ, England
Penguin Putnam Inc., 375 Hudson Street, New York, New York 10014, USA
Penguin Books Australia Ltd, Ringwood, Victoria, Australia
Penguin Books Canada Ltd, 10 Alcorn Avenue, Toronto, Ontario, Canada M4V 3B2
Penguin Books (NZ) Ltd, 182–190 Wairau Road, Auckland 10, New Zealand

Penguin Books Ltd, Registered Offices: Harmondsworth, Middlesex, England

First published in the USA by John Wiley & Sons 1998
First published in Great Britain by Allen Lane The Penguin Press 1998
5 7 9 10 8 6 4

Printed and bound in Great Britain, by The Bath Press, Bath

A CIP catalogue record for this book is available from the British Library

ISBN 0–713–99161–5

The waves of the sea, the little ripples on the shore, the sweeping curve of the sandy bay between the head-lands, the outline of the hills, the shape of the clouds, all these are so many riddles of form, so many problems of morphology, and all of them the physicist can more or less easily read and adequately solve. . . . Nor is it otherwise with the material forms of living things. Cell and tissue, shell and bone, leaf and flower, are so many portions of matter, and it is in obedience to the laws of physics that their particles have been moved, moulded, and conformed.

D'Arcy Thompson, *On Growth and Form,* 1917

We wish to suggest a structure for the salt of deoxyribose nucleic acid (DNA). This structure has novel features which are of considerable biological interest.

Francis Crick and James Watson, *Nature,* 1953

Contents

Preface

 What is life? Where did it come from? Why is the world of living creatures so different from the inorganic world?

How special *is* life, anyway?

Questions such as these have vexed the human race since the dawn of science—indeed, they were being asked long before science as such got going at all. For thousands of years, the debate has been raging. The main bone of contention is the apparently special nature of living things. Organisms are versatile, flexible, and unpredictable; they can make choices, they react to their environment; above all, they reproduce. The organic world seems utterly different from the inorganic world: Biology and physics appear to be poles apart.

Or so it has seemed.

Philosophers and scientists have argued about whether it is possible, even in principle, to understand life in the same kind of way that we understand the motion of the planets or changes in the weather. It used to be thought that living creatures were made from a special kind of matter, fundamentally different from the matter from which air, oceans, mountains, or continents are formed. On the other hand, maybe life is just a special configuration of the usual inorganic building blocks. The origins of life have also been hotly disputed. Some people think that life is so special that it can only have arisen through the intervention of a supernatural being. Others maintain that in a sufficiently rich environment, given enough time, ordinary matter will *necessarily* configure itself in ever more complex ways, until life emerges spontaneously. These views are the extremes; there are hundreds of variations in between.

Until the middle of the twentieth century, it was totally unclear whether life had any kind of inorganic basis. The discovery of the *first* secret of life, the molecular structure of DNA (deoxyribonucleic acid), solved that particular riddle. Life is a form of chemistry—but chemistry unlike any that ever graced a test tube, chemistry so complex that it makes an industrial city look like a village. Inside every living creature on Earth—and we know of none off it—a complex molecular code, a *Book of Life,* prescribes the creature's form, growth, development, and behavior. Our fate is written in our genes.

Without any question, this discovery was one of the most significant ever made. It irrevocably changed our views about the living world; it opened up entirely new ways to unravel many of life's secrets—but not all of them.

Some secrets lie deeper than the genetic code. Genes are fundamental to earthly life, but their role in determining form and behavior tends to be overstated—especially in the media. Genes are not like engineering blueprints; they are more like recipes in a cookbook. They tell us what ingredients to use, in what quantities, and in what order—but they do not provide a complete, accurate plan of the final result. Every cook knows that a recipe is not the same as a meal: Between the cook and the dining table lie the intricacies of ovens, grills, pots and pans, seasoning to taste, and the maddeningly obtuse behavior of ingredients. Last week, the recipe for bread worked perfectly, but this week's bread is as flat as a pancake. You won't find out why by studying the recipe, or the oven, or even both; you must also take account of the physical and chemical laws that govern water, bicarbonate of soda, hot air, and sticky dough—and a thousand other things.

In trying to understand life, however, it is *so* tempting just to look at life's recipe book—its DNA code sequences. DNA is neat and tidy; organisms are messy. DNA can be captured by little more than a list of symbols; the laws of physics require sophisticated mathematics even to state them. Also, the amazing growth in our understanding of genetics has opened up so many fruitful lines of research that it will take decades to follow up the most obvious ones, let alone the more elusive ones.

As a consequence, we are in danger of losing sight of an important fact: There is more to life than genes. That is, life operates within the rich texture of the physical universe and its deep laws, patterns, forms, structures, processes, and systems. Genes do their work within the context of physical laws, and if unaided physics or chemistry can accomplish a task, then the genes can safely leave them to it. Genes nudge the physical universe in specific directions, to choose *this* chemical, *this* pattern, *this* process, rather than that one, but the mathematical laws of physics and chemistry control the growing organism's response to its genetic instructions.

The mathematical control of the growing organism is the *other* secret—the second secret, if you will—of life. Without it, we will never solve the deeper mysteries of the living world—for life is a partnership between genes and mathematics, and we must take proper account of the role of *both* partners. This cognizance of both secrets has run like a shining thread through the history of the biological sciences—but it has attracted the mavericks, not the mainstream scientists. Instead of thinking the way most biologists think, these mavericks have been taking a much different approach to biology by thinking the way that most physical scientists and mathematicians think. This difference in working philosophy is the main reason why understanding of the deeper aspects of life has been left to the mavericks.

One of the great mavericks was a mathematically trained zoologist named D'Arcy Wentworth Thompson. Thompson was born in Edinburgh in 1860 and spent most of his life based in Scotland, first as professor of biology at Dundee and then as senior professor of natural history at St. Andrews. He was knighted in 1937 and died in 1948, six years after his masterpiece, *On Growth and Form,* went into its second edition. His pioneering book pointed out the success of the physical sciences in understanding nature's patterns, and it advocated a similar approach to the biological sciences. The heart of Thompson's argument was the existence of strong mathematical patterns in the organic world—the spiral shapes of shells, the curious numerology of plants, the black-and-white stripes of zebras, the fluid forms of jellyfish. He didn't just catalog those patterns; rather, he tried to find the deep *physical* principles that explained them.

Not surprisingly, Thompson's book now looks quaint and old-fashioned—certainly on the surface. In the 80 years since the maverick zoologist first put his thoughts into print, biology has changed beyond all recognition. Its whole focus has shifted—away from entire organisms and toward ever smaller features of the living world: cells, membranes, and molecules. To a modern biologist, Thompson's discussions seem naïve and outdated. It is therefore very easy to dismiss his central contention that many aspects of life are founded in the laws of physics and can best be understood by invoking the pattern-seeking science of mathematics.

Behind all that old-fashioned physics and biology, however, lies a deep truth. The discovery of DNA does not solve what Thompson called the "riddles of form." It changes the background against which those riddles must be solved—but it hasn't yet provided answers. Further, unless scientists match the tremendous progress they have made in molecular genetics by making comparable progress in understanding the mathematical basis of life, it never will. In this regard, Thompson's underlying viewpoint is as fresh and as valid now as it ever was.

DNA is not *the* secret of life—as Francis Crick excitedly and somewhat prematurely announced to the bemused clientele of a Cambridge (England) tavern more than forty years ago. It is an essential secret, but not the only one. A mathematician would say that DNA is necessary but not sufficient. Crick and Watson propelled DNA to center-stage stardom, but Thompson had his sights set on a deeper secret still: the behind-the-scenes operation of nature's fundamental laws, offstage, where life's *other* secret lurks.

If we are going to find this second secret, we must begin by recognizing that biology is not the only science that has undergone a revolution since Thompson's day. Physics and mathematics have also changed beyond recognition, becoming more powerful, more general, more flexible, and a lot closer to the intricacies of life. These advances offer radical new opportunities for uniting the biological and mathematical worldviews, at a time when there is a renewed and urgent need for just such a unification.

I predict—and I am by no means alone—that one of the most exciting growth areas of twenty-first-century science will be biomathematics. The next century will witness an explosion of new mathematical concepts, of new *kinds* of mathematics, brought into being by the need to understand the patterns of the living world. Those new ideas will interact with the biological and physical sciences in totally new ways. They will—if they are successful—provide a deep understanding of that strange phenomenon that we call "life": one in which its astonishing abilities are seen to flow *inevitably* from the underlying richness, and the mathematical elegance, of our universe.

The first glimmerings of this new fusion of sciences can already be seen. Mathematics—new, vital, creative mathematics—now informs our understanding of life at every level from DNA to rain forests, from viruses to flocks of birds, from the origins of the first self-copying molecule to the stately and unstoppable march of evolution. Our current mathematical understanding of biology, admittedly, is fragmented, piecemeal, and open to dispute—just as it is for any new science. Incomplete or ill conceived as these fragments may turn out to be, they are already utterly fascinating—especially to anyone with the imagination to see where they might lead. Or so I hope to convince you.

Here is the appropriate place to express my thanks to various people and institutions. The University of Warwick provided a year's sabbatical leave, spent mostly in the United States, Australia, and New Zealand. I am grateful to my friend and colleague Marty Golubitsky and to the Mathematics Department at the University of Houston for hospitality, an apartment, and a car. John and Vivien Casti kindly allowed my wife, Avril, and me the use of their beautiful, peaceful, and relaxing guest house in Santa Fe, New Mexico; the Santa Fe Institute managed to conjure up desk space when there wasn't any. My editor Emily Loose, at Wiley, provided endless patience, forbearance, and invaluable help, ranging from finding pictures and relaxing deadlines to pointing out what the book really ought to be about and insisting that I tell the story that way. Ravi Mirchandani encouraged the original idea and helped me develop it into something readable. Finally, thanks to Jack Cohen for teaching me biology, especially the bits you don't find in textbooks. Jack also took on the task of reading the entire manuscript (over Christmas), eliminated numerous infelicities and confusions, and explained the many exceptions to my simplified descriptions. It is not his fault that I have sometimes *left* them simplified, nonetheless: One thing he has taught me is that in real biology, there are exceptions to almost anything you say. In order to maintain a comprehensible story line, you can tell the truth—but not always the whole truth.

Coventry, England; Edinburgh, Scotland; Houston, Texas; Minneapolis, Minnesota; Santa Fe, New Mexico; Singapore; Kaanapali, Hawaii; and various cities in Australia and New Zealand.

October 1996–April 1997

1

What Is Life?

The zoologist or morphologist has been slow, where the physiologist has long been eager, to invoke the aid of the physical or mathematical sciences; and the reasons for this difference lie deep. . . . Even now the zoologist has scarce begun to dream of defining in mathematical language even the simplest of forms.

D'Arcy Thompson, *On Growth and Form*, Chapter I

In most respects, we live on a very ordinary planet. As astronomers focus their telescopes deeper and deeper into the universe, there appears to be nothing very special about our Earth and its Sun; the materials from which Earth is formed; its age; its orbit around its parent star; the size, shape, color, and temperature of that star; or the location of Earth and its star in the surrounding galaxy. Yet in one respect, Earth is quite extraordinary. It is extraordinary because we live on it— not in the sense that we might have lived on a *different* planet, but rather in the sense that our homely Earth is exactly the kind of planet that can be inhabited by creatures like us: living creatures.

If Earth were the only planet we knew about, we would probably imagine that life must be very common in the universe—for on our home planet, we find life almost everywhere. Indeed, it is hard *not* to find life. We find it in the middle of the most inhospitable deserts, in the wilderness of Death Valley, at the bottom of the deepest oceans, and in the sulphurous vents of volcanoes. Very recently, scientists have discovered rudimentary life-forms—bacteria—deep underground, thousands of meters beneath our feet. These bacteria seem to have been there for billions of years, living, reproducing, and dying . . . they may even have been the very first life-forms to appear on our planet. Perhaps instead, life began in the ocean depths, in the superheated water around undersea volcanic vents, where molten rock bubbles up from our planet's mantle and forms weird, convoluted towers and plumes of dense, black smoke.

Different as these regions are from those that *we* find congenial, it is beginning to look as if they are ideal places for life to get started. In addition, over the more comfortable parts of Earth's surface, you cannot take a step without encountering a teeming multitude of different forms of life—plants, insects, worms, mites, spiders, birds, fish, mammals, and so on.

When we use our telecopes to look at planets other than our own, however, we see little sign that life exists there now, or ever existed. The 8 other planets of our solar system, and the 10 or so now known to orbit nearby stars, may be larger than ours, smaller than ours, hotter than ours, or colder than ours. As far as we know, however, none of them—perhaps with one controversial exception, our near neighbor Mars[1]—supports life. Mercury and Venus are far too hot for life like ours to exist, and Jupiter and the rest of the outer planets are far too cold. The known planets around other stars seem even less hospitable to life.

With this realization that Earth is somehow special, the nature of life becomes a serious problem. *What* is it about Earth that is special? Could life have arisen elsewhere? What *is* life, anyway? Such questions have been answered in many ways during human history. Life has been seen as something that arises spontaneously from dust and water; as evidence for some strange, exotic kind of matter; or as the breath of God. Today, many people see life as the workings of an enormous molecular computer program, written in the language of the genes. I am going to try to convince you that as wonderful as genes are, they are not the whole answer to the question of life. More radically, I am also going to try to convince you that a full understanding of life depends on mathematics. At every level of scale, from molecules to ecosystems, we find mathematical patterns in innumerable aspects of life. It is time we put the mathematics and the biology together.

Nature already does this. A planet inhabited by living creatures differs radically from a lifeless planet. The differences are much greater than we usually imagine, and they often arise from a combination of mathematical and biological processes. For example, life has stamped indelible marks on planet Earth, in many different ways. There is a world of difference—literally—between the lifeless, cratered surface of a world such as Mercury and the rain forests of the Amazon basin. If you took a walk on Mercury, the main features you would observe would be rocks, craters, and hills (Figure 1). Deep inside a few of the craters at Mercury's poles, you might find minute traces of water. Mercury's surface is truly lifeless —not just in the sense "devoid of life," but also in the sense that nothing terribly interesting happens there. In the Amazon rain forest, on the other hand, interesting things are *always* happening. Insects are flitting from flower to flower; ants are cutting sections from leaves and carrying these in convoy to their nests, to use the leaf sections as building materials or as food. Trees are clambering laboriously to the roof of the forest, in search of the sun's light; creepers are twining their way up those trees, slowly strangling them. Tiny, jewellike frogs are swimming in rain puddles

Figure 1
The cratered surface of Mercury's Caloris Basin.

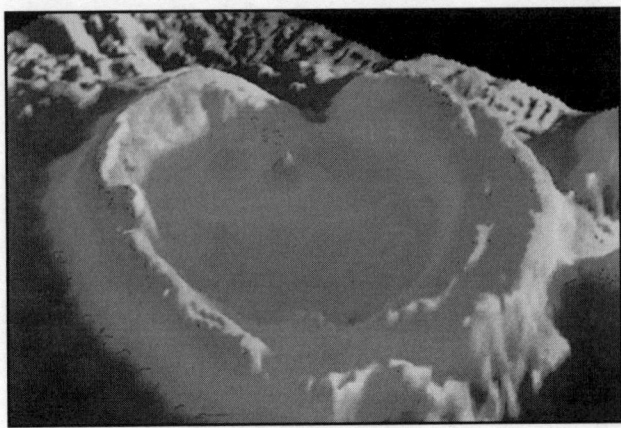

Figure 2
A computer-generated image of a volcanic caldera on the ocean floor.

30 meters (100 feet) off the ground, in the forks of trees. Furry creatures are roaming the forest floor in search of insects, worms, or seeds; snakes are slithering through rotting leaf mold, lurking in ambush.

A snake does far more interesting things than a rock does—and these distinctions are just on the surface. The difference between a living and a lifeless planet goes far deeper—again, literally. Ten kilometers (about 6 miles) below sea level, on the ocean floor, a constant rain of bits of dead organisms slowly accumulates in thick layers of sludge and slime, in a process that has gone on for billions of years. Some of those layers have turned to rock and have been uplifted in gigantic bursts of volcanic activity, to form mountain ranges far above sea level (Figure 2). Today, you can find fossil shells in the rocks of high mountain peaks—evidence that at one time, the material forming those peaks was under water.

Earth is a restless planet, and life is both a product and a cause of its restlessness. Seen from space, Earth is unusual mainly for its color: blue, spattered with ever-changing patterns of white clouds, broken by apparently changeless daubs of brown—the continents. However, if you could watch Earth for a billion years and condense your observations into a brief time-lapse movie, you would discover what earthly geologists have known for only a few decades: Those brown patches *move*. Their motion is called "continental drift." The surface of Earth consists of relatively thin plates of solid rock floating on a molten interior, and the plates jiggle about. Some move apart, and new material wells up between the plates and solidifies. Some plates slide under others, lifting the others' edges and creating mountain ranges. If you were well versed in the physics of moving rocks, you might be puzzled by how *fast* those plates move. Their motion may be painfully slow on a human scale, but on the time scale of a planet, the speed of continental drift has been much faster than you would expect from normal, inorganic geological processes. The reason seems to be that the movement of Earth's plates is lubricated by organic material from living creatures, just as oil lubricates a sticky hinge on a door.

Life has affected Earth's development in other major ways, too. Most of the oxygen in our atmosphere was probably produced by the earliest bacteria and their descendants, at a time when bacteria were the highest form of life on Earth. (I say "probably" because a rival theory, not as widely believed, holds that inorganic chemical processes in rocks may have been responsible.) So on Earth, life is not just organic decoration; it has shaped our world. No other planet that we know of is like that.

Indeed, most of the universe is empty space, a lifeless vacuum. Much of the rest is the interior of stars, where temperatures and pressures are so great that even atoms go to pieces. No star dogs can live on the Dog Star, and it is certainly conceivable that Earth is the *only* place in the universe that harbors life.

Nevertheless, the universe is big, and our knowledge of it is tiny. Most probably, some distant parts of the universe contain organized systems of matter that also deserve the epithet "life." At present, however, whether these organized systems must resemble our own living organisms or can differ considerably from our own is moot.

What *is* life? Is it a thing or a process? How did it start? Are there rules that govern its existence, its forms, its patterns, and its behaviors? If so, what *kinds* of rules?

Until 1953, biologists defined life by listing a few simple attributes, that characterized most (but perhaps not all) living organisms. Those attributes included the ability to react to the environment and the ability to reproduce. From 1953 onward, however, the answer favored by biologists became much more specific. Life is now seen as a property of special kinds of chemicals, built around the miracle molecule DNA (or, in a minority of cases, its close relative RNA, ribonucleic acid).

DNA, which stands for deoxyribonucleic acid, is like an enormously long tangled thread, formed from two strands entwined like bindweed. Along the strands are strung four special molecules, which act like the letters in an alphabet, prescribing the organism's genetic code. RNA is similar. What was the significance of 1953? It was the year in which Francis Crick and James Watson, building on experimental discoveries by

Rosalind Franklin and Maurice Wilkes, worked out the molecular structure of DNA—the famous double helix (Figure 3). Within a few years, it became clear that DNA can store information about the development of an organism and that DNA can reproduce—with the aid of a suite of other more or less complicated molecules. Moreover, on our home world, *all* living organisms exist by virtue of those properties of DNA and RNA.

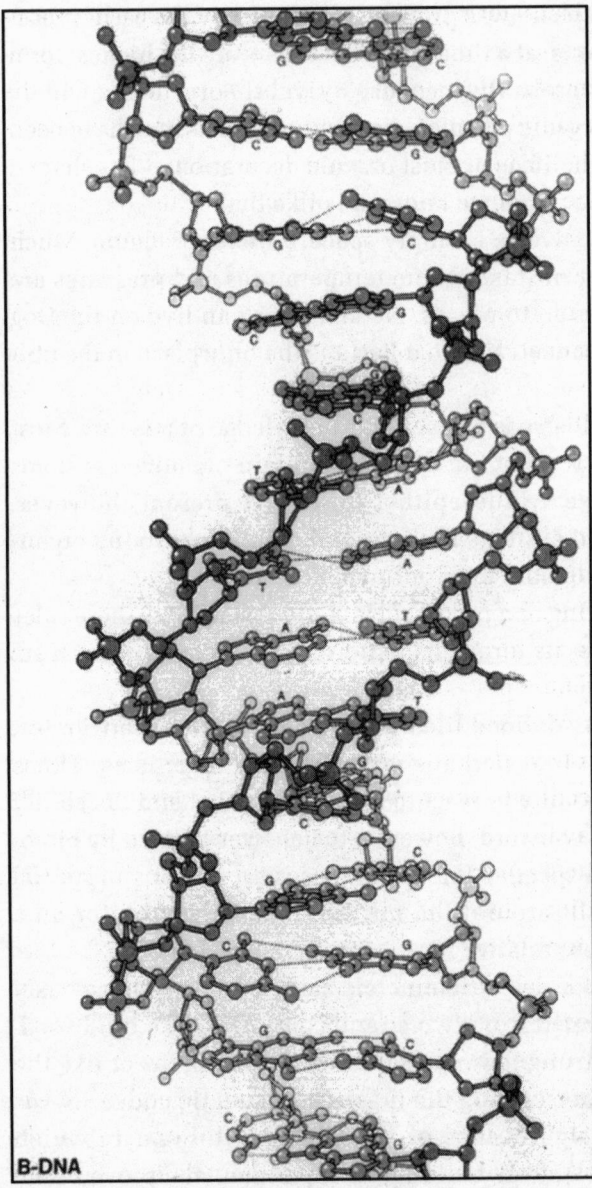

B-DNA

Figure 3
Model of the molecular structure of the commonest "B" form of DNA.

The discovery of DNA's key role in earthbound life was one of the most significant scientific advances of this—perhaps any—century. It did have one down side, however: It focused attention so strongly on the chemistry of life, and on the code incorporated into an organism's DNA, that other deep and important questions fell by the wayside.

Perhaps the biggest of these questions is whether life fundamentally differs from inorganic matter and processes, or whether it is just a rather striking example of entirely normal inorganic processes that have gathered momentum, like a runaway train, until their conseqences have dominated the history of our planet. It is too easy to be overwhelmed by the amazing *consequences* of living processes and to imagine that the processes themselves must be as amazing as their consequences.

How different are the inorganic and the organic worlds, *really?* It is on precisely this question that current controversial claims for life on Mars rest. In 1996, a team of NASA scientists announced that they had found traces of life in a meteorite that had been torn from the surface of Mars by an impacting comet, had traveled to Earth, and had crashed in the frozen Antarctic wastes. Trapped in that ancient rock were molecules normally associated with living creatures, strange formations normally associated with living creatures, and tiny bacteria-like shapes that might even be fossils of living creatures. The main objections to the inference that life once existed on Mars all boiled down to the same thing: Every one of those signs that might indicate life could also have arisen by purely inorganic processes. Thus, the question of life on Mars remains open— because life and nonlife are so close that the difference cannot reliably be detected from such remnants.

Another question is whether DNA is essential to life. Is DNA a feature of any system that obeys the old-fashioned list of life's attributes, or is DNA merely a local accident—the way our planet achieves those attributes, but not the only way to do so? Of course, there is no reason to expect a simple black-and-white answer to these questions because they raise a very deep issue. We know that our universe obeys simple low-level rules—laws of nature, including rules for subatomic particles and for space and time. We also know that life behaves in ways that do not seem to be built explicitly into those rules. Life is flexible; life is free; life seems to transcend the rigidity of its physical origins.

This kind of transcendence is called "emergence." Emergence is not the absence of causality; rather, it is a web of causality so intricate that the human mind cannot grasp it. We cannot understand how a frog works by listing the movements of every atom in it. In some sense, the atoms are the cause of the frog's behavior—but that's a totally useless way to approach frog biology. In order to understand the deeper significance of life, we desperately need an effective theory of emergent features. We need to understand how complicated rule-based systems manage to produce robust high-level features that themselves obey their own kind of emergent rules. We also need to develop an understanding of the high-level commonalities among different systems—the underlying unities, the common features that do not depend on particular interpretations and realizations, the features that life exploits to make itself possible, the universals, God's book of patterns.

Historically, we have studied the inorganic and organic worlds in very different ways. Our understanding of the inorganic world relies almost completely on mathematics. People such as Galileo Galilei, Johannes Kepler, and, above all, Isaac Newton opened up our understanding of the inorganic world by reducing it to mathematical equations. The equations represented natural processes; their solutions told us how those processes behaved. For example, one system of equations described how matter moves when pushed by a force. Another described the force exerted by gravity. Put the two together, and you can calculate the motion of the entire solar system and—at least in principle—that of the entire universe.

Modern physics has pursued this line of attack with breathtaking single-mindedness, discovering many other mathematical laws that underlie the physical universe.[2] One of the greatest triumphs of modern physics is another system of equations, which determines the behavior of matter on very tiny scales of space and time. This system, called "quantum mechanics," depicts a world quite alien to everyday experience. The quantum world is a swirling fog of probabilities, in which chance is a fundamental feature of existence, and where matter has a degree of fuzziness so that it may be doing several different things at the same time—until you try to find out *which* thing it is doing. At this point, the quantum nature of reality suddenly collapses to doing just one thing or the other, at random. Despite the influence of random events, all of this strange behavior at small scales obeys strict mathematical equations.[3]

This mathematical approach has worked wonderfully well for physics—but what of biology? Life seems not to be like these mathematical systems, whether Newtonian or quantum. Biology is more closely typified by the Harvard law of animal behavior, which says that experimental animals, under carefully controlled laboratory conditions, do what they damned well please. Their behavior is neither rigid nor random. Physics is neat and tidy, obeying mathematical equations; biology is organic and messy, obeying just its own whims and fancies.

Despite this messiness of biology, I still ask whether there could be mathematical laws that underlie the behavior of living organisms, analogous to the mathematical laws that underlie the behavior of inorganic matter. Is there a mathematics of life? There is a trivial way to answer my question, which I mention now in order to rule it out. If we accept that living creatures are made of ordinary matter, obeying the laws that govern ordinary matter, then there must exist some level of description on which an organism can be descibed as just ordinary physics and chemistry. After all, a tiger is just a collection of atoms. Write down the equations for those atoms, specify exactly what they are, where they are . . . add in the atoms in the forest where the tiger lives and those of all the other inhabitants of the forest—the bee that stings the tiger's nose, the louse that crawls through its hair, the grass upon which it treads—and now all you have to do is solve the resulting massive system of equations. In some sense, you've reduced the tiger to mathematics.

You can't do it in practice, of course—there are too many atoms. Besides, nobody could solve the equations anyway, not even with a supercomputer. Even if a complete atomic description were feasible, however, this approach would not provide an effective mathematical description of a tiger. It's too complicated, too arbitrary—and, worst of all, it would tell us nothing useful about tigers—not even that they have stripes.

Science is *not* about devising hugely complex descriptions of the world. It is about devising descriptions that illuminate the world and make it comprehensible. The reason that Newton's law of gravity is important is not because it describes the movement of every particle in the solar system. It is important because it opens up the possibility of *simple* models of the solar system that are comprehensible to human beings—models with 2 bodies, or 3, or 20, but not quadrillions. Similarly, any kind of equation for life has to be comprehensible, as well as to correspond to how organisms function—*on some level of description.*

That requirement for comprehensibility rules out trivial answers, but it doesn't of itself provide any nontrivial ones. Life is a big puzzle. Not so long ago, life seemed *so* different from nonlife that people assumed it had to be made of different *stuff.* This view, known as "vitalism," was attractive because it explained at once why life was so unusual. Of course it must be if the stuff that life is made from is unusual! Vitalism also fitted comfortably into a religious agenda: Science could safely be restricted to ordinary matter, but only the deity could have access to the ineffable material of life. Now, however, very few people agree with vitalism—not even religiously minded people.

Why did this vitalism lose favor? The reason is that when you look at the material constituents of living creatures, they seem to be exactly the same as those of inorganic matter. There is no special *kind* of matter, no spark of life, that distinguishes the living from the nonliving. It is of course conceivable that life differs from nonlife by virtue of a type of matter that cannot be detected by science—but that's a rather feeble kind of special pleading, and it just doesn't convince most people. Instead, it now seems very clear that the difference between life and nonlife is one of organization, not of component parts.

For an analogy, think of a car. A car can *move,* but a collection of automobile parts and a can of gasoline cannot. What is the difference? It is how the parts are organized. Fit the various parts together in just the right way, pour in the gasoline, push the starter, and suddenly you have a device that moves, instead of a heap of motionless junk. Take apart the car again, and every component is still there—but the *motion* isn't.

Which explanation makes more sense: one based on motion as an invisible and ineffable *substance,* or one in which it is a process that can occur only when the parts are suitably organized? I know where I'd put my money, and so do you. None of this *explains* life, but it does help us to see what kind of thing life is: a process, not a substance.

We still don't have a comprehensive understanding of that process. In many ways, we don't really know what life is. Is DNA a universal necessity or just a parochial accident—the way life got started here, but not the only way it can start? An awful lot hinges on how well life on Earth represents life in general. Suppose, for the sake of argument, that elsewhere in the universe there exist entities that deserve the description "life." Must they be "life as we know it"? Is DNA the *only* molecular basis for forms of life, or could the whole game be played in a different way?

There are many possible degrees of difference. Some are so small that they are unlikely to be significant. For example, our particular genetic code, used to turn DNA sequences into proteins, is somewhat arbitrary. It is easy to imagine new forms of life that use DNA in exactly the same way as life does on Earth, but with a different code. Such life-forms would be like a new molecular dialect, rather than a new language. If we started out with a narrow definition of life, incorporating just Earth's genetic code, we would be quick to revise our definition to include such variants.

A more serious difference would arise if some totally different molecule could be used instead of DNA. Alien organisms based on such a scheme would still have genomes, but the chemistry would be very different in detail. Alien biology would completely rewrite the earthly textbooks (Figure 4). In the abstract, however, there would still be a strong resemblance to life as it exists on Earth, and again there would probably be little resistance to extending the definition of life to encompass these new creatures. Already this points to a key feature of the concept *life:* It is how things *behave,* rather than what they are made of, that leads us to view them as living.

Figure 4
"Scattered about, some in their overturned war machines, some in the now rigid Handling Machines, and a dozen of them stark and silent and laid in a row, were the Martians—dead!"

The universe is strange, however—perhaps more strange than we can know. There is no particular reason for organisms to be made from carbon-based molecules, as they are here. Silicon is another possibility—but what of metallic, machinelike creatures? Today's technology is getting very close to being able to make a von Neumann machine, a self-reproducing robot. Such a machine could easily be given a kind of genetics, so that succeeding generations could adapt and evolve. We might begin to think that life could be built from almost anything: patterns of dust swirls in an asteroid cloud, magnetic loops in billion-degree plasma at the heart of a giant star, gravitational waveforms in the void between the galaxies. . . . Perhaps the entire universe is one single vast organism. . . . Perhaps some ways of organizing matter in the universe make life on Earth seem no more amazing than a speck of dust.

On the other hand, of course, life on Earth may be the only life there is, or it may be organized in the only possible way to do so. If so, our task is easier, and DNA really is what counts. If not, however, we have to face up to the general concept of life, not just its realization on this planet. It then becomes a major problem even to *characterize* life. Life is not a thing; it is some kind of abstract property of a system, characterized by such features as adaptability, flexibility, reproduction, self-complication, self-organization . . . and not by a specific molecular structure.

Such features then pose questions for *mathematics,* the science of structure and pattern. I believe, therefore, that we have to pursue the mathematical aspects of life on Earth, in order to be able to generalize our ideas to all possible forms of life or quasi life that do or can, in principle, exist. If we keep discovering ways in which mathematics informs biology, we open up the prospect of discovering the *deep* structure of life—what it really is. Such a discovery could eventually lead to some kind of grand unified theory of life, one that illuminates life on Earth by placing it in a far broader, and far less parochial, context.

Maybe. It's pure speculation. It won't put dollars in the pocket of an executive of a genetic-engineering corporation. On the other hand, if science abandons the deep questions, it will lose its direction, its heart, and its soul.

Popular accounts of modern genetics leave the strong impression that life depends almost exclusively on the DNA code. In particular, life is amazingly complicated because the DNA program that describes it is amazingly complicated. The single-minded focus on DNA is comfortable: Life seems to do all kinds of inexplicable things, but if those things can be written in a DNA codebook, then life's versatility is less of a surprise. A big book can hold an awful lot of instructions. So the flexibility of life boils down to a lot of contingency planning in its genetic code, and the complexity of life arises because the recipe for life is very, very, very, very long.

You know, that's a remarkably boring explanation for such an amazing thing. Indeed, the explanation of life cannot possibly be that simple. The complexity of an adult organism, such as a tiger, exceeds that of its DNA by virtually any sensible measure. The wiring diagram for a tiger's nervous system *alone* is more complex, by several orders of magnitude, than the tiger's entire DNA sequence. That sequence, complex as it is, versatile as its contingency plans may be, does not contain enough information to specify how to build a tiger's brain, let alone an entire tiger. We know this because it is possible to quantify information. If a door-to-door salesperson showed you a 20-page notebook, printed in normal-sized type, and claimed that it contains the entire *Encyclopædia Britannica,* you wouldn't need to read it to know that you were being sold a bill of goods. It's the same with DNA and tiger brains.

So how does the tiger do it? Where does it get all of its missing information? That may not be the best way to ask the question; the concept of information may be too superficial to do much more than demonstrate the existence of a problem. However, if we insist on talking in such terms, then the answer seems to be this: The missing information is supplied by the mathematical rules (the laws of physics) that govern the behavior of matter—inorganic matter—well, *any* matter. Nonetheless, when we're talking in these terms, the inorganic perspective holds sway because that viewpoint is where the existence of physical laws first came to human attention. This entire discussion brings us full circle to Thompson, maverick advocate of the biological role of physical laws and their associated mathematical patterns. To see *why* he came to that point of view, we take a step backward in time.

The year is 1917. On the continent of Europe, the British and German armies, together with allies drawn from the four ends of the earth, do battle eyeball-to-eyeball across seas of mud, which are sliced haphazardly with trenches, strewn with barbed wire, pocked with shell craters, and littered with the dying and the dead. In England, a book is published, a measured splash of calm in an ocean of madness. The book: *On Growth and Form.*[4] The author: the aforementioned D'Arcy Wentworth Thompson, an expert zoologist with a flair for mathematics and a maverick message.

The message: The organic world is just as mathematical as the inorganic world. The mathematical basis of living things, however, is more subtle, more flexible, and more deeply hidden. It is not just a matter of writing down some simple, elegant Newtonian laws of life: Thompson is no Euclid of the plant world, and there is no Thompson's equation for an animal. Nonetheless, circumstantial evidence in every corner of the living world convinces Thompson that there are real mathematical patterns in living organisms, and that the abstract principles behind these patterns can illuminate the world of living creatures in a manner that complements the more concrete concerns of traditional biology. Now all he has to do is convince everybody else.

He gave it his best shot, and in a way, he succeeded: *On Growth and Form* remains a classic and is still well regarded in many quarters. It never made it into mainstream biology, however—and for good reasons. Some of the evidence was just *too* circumstantial, some of the alleged patterns were little more than visual puns, and some of the mathematical stories just didn't hang together when faced with the real, direct evidence from biologists' laboratories. The mainstream moved on, and the maverick classic remained just that. Yet that disconcertingly unorthodox idea—that life is a partnership between biology and mathematics—refused to go away. As the mainstream moved, the idea moved with it.

Before we turn to the discoveries of the modern era, it is worth reminding ourselves of the kind of evidence that was available to Thompson nearly a hundred years ago, and of the state of biology and mathematics in his day. Only by doing that can we understand how radically both biology and mathematics have advanced in the interim—so that the question of life and its possible answers have changed equally radically.

Our point of departure is the cell. A cell is a diminutive structure, a tiny speck of protoplasm contained within a thin membrane; however, its tiny size does not make the cell simple. Every cell contains a number of distinct *organelles*, specialized structures (such as the nucleus or the mitochondrion), which carry out important tasks (such as gene management and energy production, respectively). Many living creatures—certainly, all of the more complex ones—are made from cells, a fact first discovered when microscopes were invented. The word *cell* owes its existence to a mathematician, Robert Hooke, who first noticed the cellular structure of cork in 1665. By 1674, the Dutch naturalist Antonie van Leeuwenhoek had observed bacteria 2 thousandths of a millimeter long, as well as blood cells and spermatozoa. Yet it took until 1839 for scientists to apppreciate the significance of the cell as the basic unit of the organism, through the work of people such as Theodore Schwann and Matthias Schleiden.

The human body contains roughly a trillion cells, of more than a hundred types—nerve cells, blood cells, liver cells, bone cells, muscle cells, and so on—but the body is not merely a vast cellular conglomerate. In order to function collectively as a human being, those cells must be put together in a specific, complex manner. The human brain, for instance, has abilities that the greatest supercomputer cannot match—such as being able to look at a landscape and instantly to pick out a sheepdog in a distant field. No technology on Earth can build you, nor can it build a single one of your cells. No reputable biologist would ever claim to understand anything as complex as a cell.

Cells are constantly working tiny miracles, one of the most striking of which is reproduction. A single cell can divide in two, each half forming a new complete cell, capable of reproducing again, almost indefinitely. From one cell, by repeated reproduction, nature can create huge numbers of cells, all just as complex as the original. Cells multiply by dividing. It *sounds* like mathematics—but it doesn't sound like the mathematics of elementary school arithmetic, and it isn't.

The microscopists of the nineteenth century collected pictures of cells caught in the act of dividing and tried to puzzle out how on Earth the cells did it. D'Arcy Thompson assembled some of the microscopists' pictures and noticed, despite all that complexity, clear patterns and regularities. He noticed arithmetical patterns in the arrangement of those organelles known as *chromosomes,* which contain (most of) the cell's genetic material (Figure 5). He pointed out analogies between cell division and the theories of electricity and gravitation (Figure 6); he also found analogies with chemical diffusion, in a homely experiment using ink in salt water.

The shape of a cell—just before, during, and just after division—is mathematical. In cross section, the shape is a simple curve: a circle that develops a waist, which narrows, pinches into a figure-eight shape, and breaks apart to create *two* circles. This simple shape indicated to Thompson that there must be a connection between the division of cells and the physical principles that govern the forms of soap bubbles and foams (Figure 7).

One of the big principles of physics is the idea that the inorganic world is fundamentally lazy: It generally behaves in whatever manner

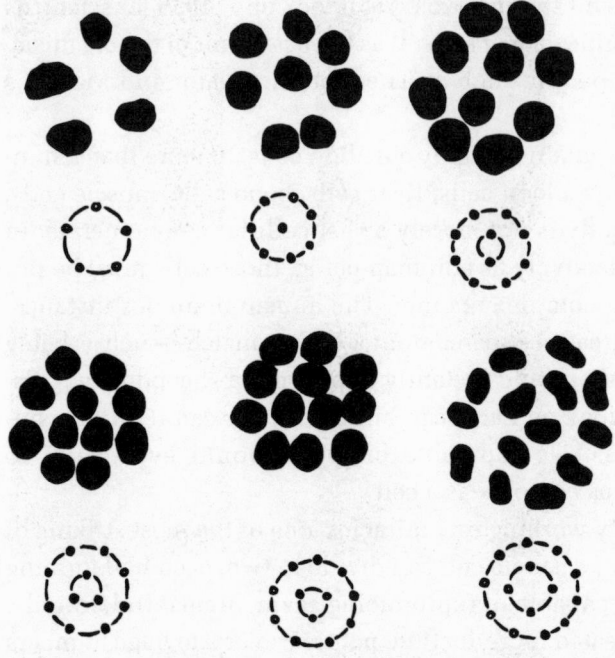

Figure 5
Arrangements of chromosomes in the equatorial plate of a dividing cell (top), and corresponding configurations of mutually repelling magnets (bottom).

requires the least energy. The energy of a soap bubble comes from the tension that holds the molecules of soap together. Think about the last time you blew up a balloon: It was hard work because you had to provide energy to set up elastic tension in the balloon's rubber surface. The formation of soap bubbles requires much the same expenditure of energy. Just as it takes more effort to blow up a bigger balloon than a smaller one, it takes relatively more energy to produce a soap bubble or film with a

Figure 6
First division in the egg of *Cerebratulus* (left), compared with the field of force between two equal electric poles (right).

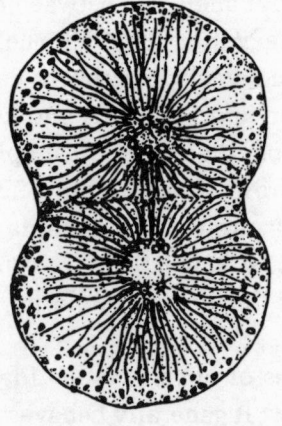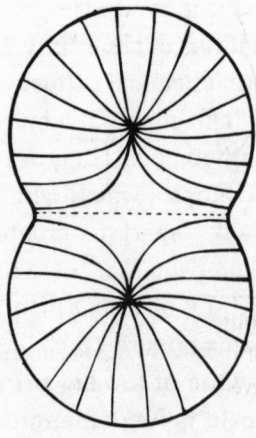

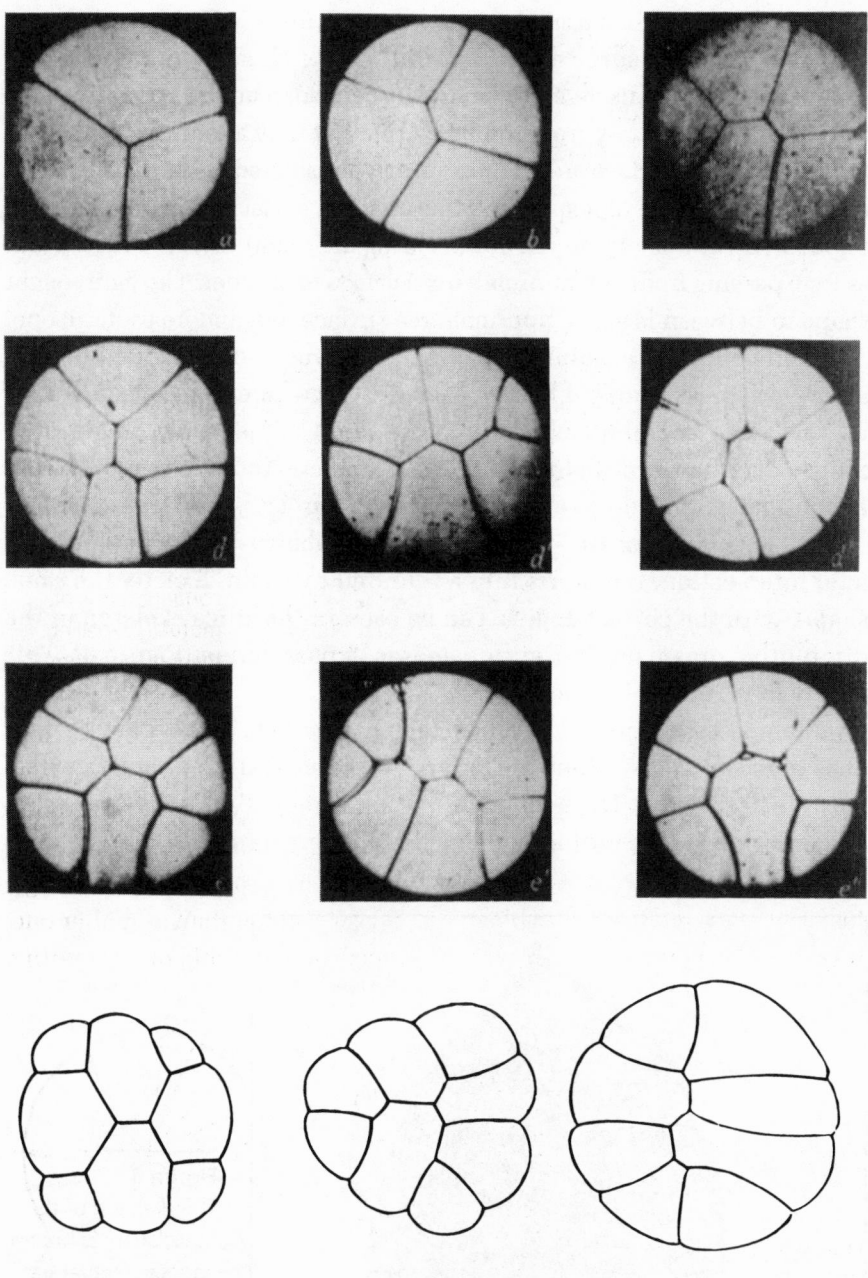

Figure 7
Partitioning of groups of soap bubbles (top), compared to patterns of division in the developing eggs of various animals (bottom).

larger surface area than a smaller one—so soap films with the least energy also have the least surface area. The blind French mathematical physicist Joseph Plateau discovered that the form of soap bubbles and films can be deduced completely from the principle that they adopt the shape that has the smallest surface area. For example, an isolated bubble is spherical because the surface of a sphere is the smallest surface area that contains a specified volume of trapped air. Dividing cells start as a sphere and end as two, passing from one minimal-area surface to another. The figure-eight shape in between is also a minimal-area surface, but a more esoteric one.

Thompson was absolutely fascinated by such surfaces, and he saw them—or at least thought he saw them—everywhere in living creatures. He saw them in cell membranes, in the shapes of jellyfish, in algae, in fungi—even in the skeletons of microscopic creatures. When four soap bubbles meet, they do so along six common surfaces, meeting each other in pairs at an angle of 109°. If a fifth, smaller bubble is trapped at the common intersection, it distorts into a rounded pyramid. Exactly the same shape, with the correct angles, can be seen in the silica skeleton of the diminutive organism *Callimitra agnesae*, a nasselarian (Figure 8). This hardly looks like coincidence.

Thompson saw other mathematical patterns, too. There were obvious ones in radiolarians, which are also microscopic marine creatures with a hard silica skeleton. The bodily scaffolding of these tiny animals displays innumerable and beautiful mathematical patterns, some of which bear a

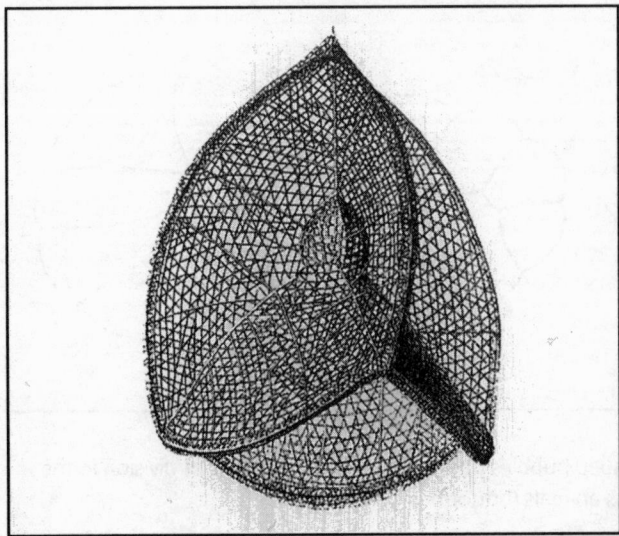

Figure 8
The skeleton of the nasselarian *Callimitra agnesae* (actual size 1.5 mm) bears a striking resemblance to surfaces formed at the junction of five soap bubbles.

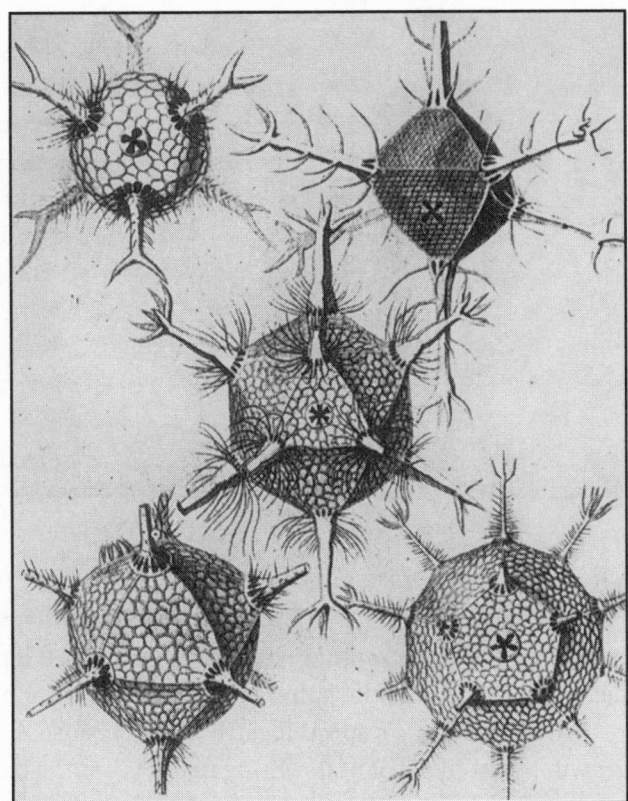

Figure 9
Radiolarian skeletons resemble regular solids.

striking resemblance to Euclid's regular solids—the octahedron, dodecahedron, and icosahedron (Figure 9). Some say the resemblance is *too* striking; the illustrator may have exaggerated their skeletal regularity. Be that as it may, these creatures are without doubt beautiful, intricate, and highly patterned. They *look* like tiny bits of living mathematics.

Another ubiquitous form of living mathematics is the spiral. All of us are familiar with the spiral shells of snails on land, and many of us also know about whelks and periwinkles in the sea. Some marine shells—bivalves such as the mussel—consist of two saucerlike shells hinged together and do not have the striking mathematical beauty of the spiral, but the majority of marine shells are based around a spiral form.

Perhaps the most elegant spiral of all can be seen in the *Nautilus*. Its shape closely resembles a curve known to mathematicians as the logarithmic (or equiangular) spiral (Figure 10). If you were to tie a stone to a string and then swing it around your head while gradually letting the string out so that its length increased by a fixed proportion when it swung through

Figure 10
Nautilus in cross
section, showing
its logarithmic spiral
shape.

a fixed angle (say, 10% longer every 30° of rotation), then the stone would follow a logarithmic spiral. So elegant is the mathematics of the logarithmic spiral that the mathematician who first understood its geometry, Jacob Bernoulli, had the spiral engraved on his tombstone.

Why is *Nautilus*'s spiral logarithmic? Because the animal's pattern of growth is just like that of the string that lets out the rotating stone: increasing by a fixed proportion as the growing shell turns through a fixed angle. In fact, the shape of the spiral shell tells us a lot about the rate of growth of the animal inside it. The shell of the fossil *Ammonite,* an extinct relative of *Nautilus,* is closer to an Archimedean spiral, in which the string is let out by a fixed *amount,* not a percentage, for a fixed angle—say 1 centimeter for every 30° of rotation. That proportion is not true near the middle, but it's a fair approximation farther out, and we can deduce from the shape that after an initial spurt of growth, the adult *Ammonite* grows much more slowly.

The mathematical nature of the biological world takes on an especially elegant and enigmatic form when we come to the plant kingdom. An entire chapter of *On Growth and Form* is devoted to the remarkable geometry and numerology of plants—the arrangement of their leaves along the stem, the curious interpenetrating spiral patterns formed in seed heads (Figure 11), and the numbers of petals. The mathematics here is really strange. Far more often than not, the structure of plants involves a curious sequence of numbers known as the Fibonacci sequence:

$$1, 2, 3, 5, 8, 13, 21, 34, 55, 89, 144, \ldots$$

Figure 11
Seed head of the
giant sunflower
Helianthus maximus,
showing Fibonacci
spirals.

The Fibonacci sequence has a nice pattern to it: Each term from 3 onward is the sum of the previous two. For example, 55 = 34 + 21. The sequence was invented, apparently off the top of his head, by Leonardo of Pisa in the year 1202. Leonardo was one of the great mathematicians, an indefatigable traveler who came across the newfangled number notations invented by the Hindus and the Arabs, as opposed to the existing Roman-numeral notations. In these systems, the same symbol could have a different meaning, depending on its position: For instance, in the Fibonacci number 55, the first 5 means "50" and the second means "5." Leonardo wrote an epoch-making book about Hindu-Arabic notation, and as a result, the Western world owes its current system of arithmetic to his act of popularization. He was given the nickname "Fibonacci" (Son of Bonaccio) by the eighteenth-century French mathematician Guillaume Libri, and it stuck so hard that most people think it comes from the twelfth or thirteenth century.

Fibonacci also devised a rabbit problem, which goes like this: Assume that at breeding season zero, we start with precisely one pair of immature rabbits, which mature for one season. Every season, each mature pair produces just one immature pair, which in turn takes one season to mature. Rabbits are immortal. How many pairs of rabbits will there be in each season? It turns out that in successive seasons, the rabbit population follows the Fibonacci sequence—and an awful lot of important mathematics has developed from this simple discovery. However, real rabbit populations don't obey Fibonacci's model, and you don't find obvious Fibonacci numbers if you go counting rabbits.

You *do* find these numbers, however, if you go counting petals, sepals, stamens, and other bits of flowers. For instance, lilies have 3 petals, buttercups have 5, delphiniums have 8, marigolds 13, asters 21, and most daisies have 34, 55, or 89—and you don't find any other numbers anything like as often. Fibonacci numbers are also hidden in sunflower-seed patterns. If you look at the illustration, you see two families of spirals, one curving clockwise, the other counterclockwise. Count how many spirals there are in each family: You'll find that the answers are both Fibonacci numbers.

Polyhedrons, spirals, Fibonacci . . . all very well, but from the point of view of today's biology, Thompson's whole discussion seems hopelessly naïve. It is *so* easy, and so tempting, to pooh-pooh the whole idea— to see his book as just a catalog of visual puns, coincidences, and chance similarities among totally different things. The rudimentary state of the biological sciences in his day didn't help: It forced him to rely on analogies, rather than experimentally verified theories; and it made his work descriptive, rather than structural. For example, although he analyzed the Fibonacci numerology of plants, he didn't explain the biological origin of the Fibonacci numbers.

Far worse, his discussion made no contact with what is now the dominant theme in modern biology: genetics. What Thompson did not know—could not possibly have known—is what Crick and Watson recognized in 1953.[5] Their epic discovery was that every cell of every living organism contains a symbolic recipe for that kind of organism, written in a molecular code. This recipe is the organism's *genome,* its total genetic makeup, written in the language of DNA. DNA plays two linked but distinct roles that make living organisms different from the rest of the physical universe. First, DNA prescribes the growth and the form—and indeed even the behavior—of the organism, using a molecular code. Second, that code occasionally *mutates*—changes because of random chemical mistakes—allowing the organism to evolve. So instead of the free-running mathematics of growth with which D'Arcy Thompson was familiar, life is perpetually tinkering with its own mathematical basis—nudging its development in preferred directions and away from undesirable ones.

How much of the mathematics of growth survives this nudging? The DNA sequence for an organism can, in principle, create *any* form or pattern—and, in principle, any form whatsoever can also evolve, provided the evolutionary change offers an advantage. Have DNA and evolution made such sweeping changes that mathematics is now irrelevant to living things? In any case, isn't life so dramatic and amazing that it *obviously* cannot be reduced to mere mathematics? Mathematics is rigid and prosaic; life is dramatic and exciting.

I want to convince you of two things: Mathematics is more dramatic and exciting than most people think, and—more importantly—today's mathematics is far closer to the flexibility of life than it is to the rigidity of Euclid. Thompson's key argument is that mathematical patterns exist in living organisms, and that these patterns must therefore have mathematical causes. Unless the bounds of coincidence are to be stretched to the breaking point, this view has to carry a lot of weight. Nonetheless, knowing what we now do about the molecular basis of life, we should avoid being naïve about what kind of mathematics is expressed in the living world. We should not expect life to display *only* mathematical patterns, and we most definitely should not expect to see pure mathematical patterns, undecorated by genetic and evolutionary tinkering.

Still, there are good reasons why a lot of the underlying mathematics should have survived such tinkering. Think of the physical universe as a source of raw materials for the development of life: not just materials in the literal sense—structures and processes, too. If you set up a physical system and let it run of its own accord, then it will do something. Often, it creates some structure—a wave, a crystal, whatever. Some of those structures can be surprisingly complicated—such as foams and dendritic (treelike) patterns in mineral deposits. All these structures are grist to biology's mill. However, life is too complex, its structure too tightly constrained, to arise easily from such origins.

Living creatures make use of the structures and processes that free-running physics provides, but those processes have to be modified and controlled before a true living organism can result. Chemical reactions, for instance, slowly run out of important ingredients and stop. Organisms solve this problem by replenishing their supply of key chemicals, a trick known as "food." No simple physical or chemical system goes looking for food, but *some* of the necessary ingredients of such a system can be found in the inorganic world. Chemicals can diffuse, spread out from their source; the farther away you are from that source, the weaker the concentration of the chemical is. Reversing that process, anything looking for that chemical can climb the gradient, moving in whichever direction increases the concentration. So the physical universe supplies a trick that *could* be used by an organism to look for food.

Genes come into the picture by making sure that this trick *is* used. Genes add a lot of flexibility to growth and form because they control and select the physical patterns that the organism needs. With the help of evolution, any genetic tinkering that works—does something new and useful—gradually becomes more and more sophisticated. Nevertheless, *most* of the time, nature employs relatively uncontrived mathematical patterns. You find an awful lot of spots, stripes, patches, blobs, and other patterns of a type that are bread and butter to mathematicians. The surface markings of tropical fish, despite their intricacy, are similar to those produced by entirely straightforward mathematical processes, and the same goes for markings on seashells, insects, and mammals. It is true that in birds, the forms and patterns do indeed become considerably more exotic; birds of paradise, for instance, are famous for their fringes, curlicues, spikes, and crests. Also, the basic unit of bird pattern is the feather, which is not a shape traditionally employed by mathematicians. Bird patterns can even change, often in dramatic ways, when the bird changes its position and alters the *register* of the feathers—the way they sit next to each other.

Nevertheless, the patterns found on birds are made from mathematical *ingredients*. The pattern-making mechanism fits those ingredients together in rather arbitrary ways—but that's because evolutionary pressures on markings seem to have been very strong for birds. An exotically marked bird is rather like what would be produced by a mad mathematician with a big pattern book and a pair of scissors—a collage of different mathematical forms, rather than a unified whole. Nonetheless, the mathematics is still present.

Although genetics and evolution are very flexible, they cannot actually do *anything*. They can find clever ways to harness physical laws to counterintuitive ends, such as reproduction, but they can't *break* those laws. Physics imposes constraints on what biology can do. For example, it is well established that the form of birds' tails depends heavily on aerodynamics: A tail that stops a bird from flying is a liability.[6] What bird genes can accomplish regarding the shape of tails is constrained by physics—and without understanding the physics, you won't understand *why* the genes have evolved to build the kinds of tails that they do.

Consider also the hemoglobin molecule, which picks up oxygen from our lungs, carries it in our bloodstream, and releases it where it is needed. Hemoglobin is a highly complex molecule, a protein made by sticking together a large number of units known as "amino acids." A complete hemoglobin molecule acts rather like a pair of pincers: It snaps shut around an oxygen molecule, holds it in its jaws, and opens to release it again sometime later. These abilities depend very sensitively on the precise shape of the hemoglobin molecule, and on its ability to click into slightly different positions—open and shut, so to speak. Genes prescribe the amino acids that must be put together to make a hemoglobin molecule, but they do not prescribe its actual shape. Indeed the protein folding problem—to predict the three-dimensional geometry of a protein molecule from its sequence of amino acids—lies at the very frontiers of today's science, and we do not understand it at all well. We do understand, however, that the shape of a protein molecule is controlled by more than its genetic code. The shape is a consequence of deep laws of physics and chemistry, which are expressed in mathematical form. So, ultimately, the shape of hemoglobin depends on mathematics.

Genes are not the laws of life; they are what the laws use to operate. As a loose analogy, the current state of the solar system is determined by two things: mathematical laws of motion and gravitation, *plus* a list of initial conditions that tell us where all the planets are at some chosen instant. Plug the initial conditions into the laws, and all of the intricacies of the solar system follow. There is a strong tendency to think of genes as constituting the laws of biological development, but that's not so; their role is much closer to that of initial conditions. In other words, genes are not *the* key to life. They are *a* key, and an enormously important one, but behind them lies something much deeper. There must be more fundamental theories, the true laws of biology, the mathematical rules into which the genetic code is plugged.

Physics provides a range of patterns and structures that are available for no extra expenditure of energy. Both evolution and genetics can, each in its own way, modify those structures and patterns, fine-tune them, and put them together in ways that would not be natural for raw physics;

nevertheless, the mathematical patterns provide building blocks and a place from which genetics and evolution can start. Moreover, if it so happens that these freely available forms do their job effectively without modification, then evolution will select them, and genetics will respect them. In this manner, these structures and patterns have been built into the very fabric of life.

A final plea: Don't expect any mathematical model to explain every tiny detail of the rich panoply of life. Even in physics, nobody expects mathematics to be *that* effective. Mathematics tells us why Mars is roughly spherical, how it rotates on its axis, and how it moves around the Sun. It even gives us an understanding of how a meteor impact can create a crater. Nonetheless, it would be silly to expect mathematics to explain every single feature on a map of the entire planet, all in one go. Let's not make mathematics jump through hoops in biology when it can't even jump through them in the physical sciences.

Having understood how, and to what extent, mathematics might help to explain living creatures, we are now in a position to examine the new evidence that it *does so*. That evidence goes well beyond anything available to D'Arcy Thompson: It concerns not just the growth and form of organisms, but also their molecular bases, their microstructures, their control systems, their movements, their patterns, their behaviors, their means of communication, their relationships with each other and with their environment, and the historical paths by which they evolved. There is mathematics at every level of life; to see it, you need only look.

So let's do just that.

2

Before Life Began

Crystals lie outside the province of this book; yet
snow-crystals, and all the rest besides, have much
to teach us about the variety, the beauty and the very
nature of form.

D'Arcy Thompson, *On Growth and Form,* Chapter IX

An amoeba, seen with the naked eye, is less than impressive—a tiny speck, assuming it is large enough to be visible at all. It looks like a grain of sand, and you expect it to be just about as interesting. However, when you look at an amoeba through a moderately powerful microscope, you suddenly realize just how remarkable life is. For an amoeba is not like a grain of sand; it doesn't have a fixed form. It moves—purposefully.

It's difficult not to use such a word, unscientific as it may be. You can't help receiving the strongest impression that the amoeba knows where it wants to go and what it has to do to get there. Nearby is a particle of food; you can see it through your microscope. The amoeba knows it's there too, for the amoeba begins to reach for it. Its surface starts to change, forming a bump, which lengthens rapidly, heading unerringly toward the food particle. The amoeba has extended a *pseudopod,* a flexible armlike protuberance. Within the pseudopod are tiny granules of some kind of material, circulating through the pseudopod, flowing toward its growing tip.

Then, suddenly, the creature changes its mind. The flow of granules reverses, the pseudopod shrinks. The amoeba moves off in a different direction altogether, for no apparent reason.

You remove your eye from the microscope and inspect the amoeba again without optical aids. Once more, it just looks like a grain of sand—but you will never underestimate the abilities of a speck of *living* matter again.

Today, our planet teems with life. Yet 4 billion years ago, it was a barren wasteland of semimolten rock, steaming sulphur vents, and lifeless mineral-laden seas. Somehow, from these unpromising beginnings, the rich complexity of life emerged. At first sight, it is difficult to see how this could have happened—the gap between life as we know it and the sterile world of inorganic matter just seems too great. Sand blowing about on a beach moves, yes, and it even moves in patterns, but it is obviously just responding to the wind, and doing so in a very simple manner. The wind pushes it, and the sand goes in whatever direction it is pushed. Living creatures aren't like that: Witness our amoeba and its apparent sense of purpose, its freedom of movement.

I'm not suggesting that the amoeba knows where it wants to go. All I'm saying is that to us, it *looks* as if it's choosing to move in whichever direction it wishes; it does not appear to be subject to the whims of its environment in the same manner that a grain of sand is. And that's just an amoeba. What about a cheetah?

Our everyday world seems to contain two types of matter: living and nonliving. Nonliving matter does pretty simple things; living matter is inconceivably complicated. It is hardly surprising that it took thousands of years even to begin to understand life and to realize that life might have inorganic beginnings. The gap between life and nonlife seems just too large to bridge.

But is it? My immediate task is to convince you that the assumption of an unbridgeable gap is a misconception, based on too limited a view of the inorganic world. Nonliving physics is capable of far more intricate behavior than it has traditionally been credited with; rigid mathematical schemes can produce astonishingly flexible results. The gap between life and nonlife may well be a nongap: Instead, there is a continuous spectrum of behavior, rigid at one end, gloriously alive at the other—but with no obvious boundary in between. At what stage physics and mathematics cease, and biology takes over, is a matter of taste.

Not so many years ago, the gap seemed much clearer. Biologists used to think that there was a huge gulf between the complex chemistry of living creatures and the far simpler chemistry of unaided molecules. When explaining the emergence of life, they found it necessary to erect a complicated logical structure of supporting chemistry. The chemistry became more and more complicated the closer it got to the chemistry of life. Today, we understand that unaided chemistry is capable of producing very intricate structures and processes in its own right.

I say "chemistry," but what we are really discussing here is molecular architecture, not the kind of thing you see in laboratory test tubes. In test-tube chemistry, everything is mixed together more or less uniformly, so that all the interesting spatial structure—different things happening in different places—is lost. Well, you wouldn't get anything terribly interesting from living organisms, either, if you ground them into a homogeneous paste; so already we see that the apparent gulf between inorganic and organic chemistry might be smaller than we thought. In fact, the inorganic world is an almost inexhaustible source of patterns, many of which are astonishingly elaborate. These patterns include snowflakes, waves, crystals, bubbles, droplets, and mineral dendrites . . . and all of them are consequences of mathematical laws.

Life may *look* different from mathematics—but that's because we have the wrong idea about mathematics. Most people have some sort of mental image of what biologists, or physicists, or astronomers—or indeed bank managers—do. They study living creatures; they carry out huge, expensive experiments on the fundamental constituents of matter; they look through telescopes at the stars and planets; or they lend money to people. I'm not worried here about the extent to which such images are correct—they capture some of the essential spirit of those enterprises, even though they are actually rather wide of the mark when it comes to details. What concerns me is that when we think of mathematics, the only mental image that most of us have is what we did at school, and we tend to assume that this is all the mathematics that exists.

Not so. Mathematics is not a long-dead subject preserved in dusty tomes, in which all the questions have been solved and all the answers are listed at the back of the book. It is a vibrant, lively, ever-growing subject: Indeed, more new mathematics is being created today than ever before. Further, this new mathematics is not just ever-more-complicated answers to bigger and bigger sums. It lies on a far higher conceptual level. Mathematics is the study of patterns, regularities, rules, and their consequences—the science of significant form—and nowhere is form more significant than in biology.

This view of mathematics may sound rather abstruse, but it actually makes the mathematics of life more interesting and easier to understand than the prosaic techniques taught at school. A fair analogy is the difference between practicing scales on a musical instrument (school mathematics) and composing (creative mathematics). The mathematics that may one day provide an understanding of life in *all* of its aspects, its deep generalities, will be the creative kind, not the prosaic.

Instead of arguing this contention in the abstract, I'll let the inorganic world speak for itself, by taking a look at some representative patterns. I'll start with some familiar and relatively simple examples and then move on to more esoteric and more complex ones. My intention is to demonstrate just what the inorganic world is capable of, and to begin to bridge the apparent gap between the worlds of nonliving physics and living biology.

I'll start with a well-known pattern in the inorganic world, one known to Kepler four centuries ago—yet one with secrets so subtle that they have only recently yielded to scientific investigation. To about half the people in the world, those who live in cold climates, few things could be more familiar than the snowflake. (To the other half, few things could be *less* familiar.) Even those of us who regularly encounter snowflakes, however, seldom take a close look at them. We *know,* because we have seen pictures or have been told, that snowflakes are tiny crystals of ice, and we may well recall that they are rather beautiful when magnified, having a delicate branching structure with six-sided symmetry. In 1880, an American named Wilson Bentley made photographs of snowflakes through a microscope, collecting the tiny crystals in an open-windowed hut facing into the howling winter gales. Bentley published a book containing 2,500 photos of snowflakes. The branching structure varied from one ice crystal to another—no two snowflakes in his collection were the same. Nonetheless, the great majority of them had a very striking form: perfect sixfold symmetry. They were elaborate, tiny hexagons—not the simple geometric hexagon of the mathematician, with its six straight sides: no, the form was intricate and beautiful, like a fern made of glass. In one key feature, however, the snowflakes resembled the mathematician's hexagon: They were like six identical ferns, arranged in a neat bouquet at 60-degree angles to each other. The simplest way to convince yourself firsthand that this beauty really does occur is to go out into gently falling snow, wearing a dark coat and carrying a magnifying glass. (Bear in mind that real snowflakes are often imperfect, perhaps because some pieces break off or melt.)

The first serious recorded attempt to explain the form of snowflakes was made in 1611 by Kepler, who is best known for his three laws of planetary motion. Kepler was an inveterate pattern seeker, interested in any aspect of reality that exhibited traces of mathematical regularity. He wrote a wonderful booklet, *The Six-Cornered Snowflake*,[1] as a New Year's gift to his sponsor. In it, he traced the sixfold symmetry of snowflakes to their atomic structure. The theory that matter is made from extremely tiny particles—atoms—goes back to the ancient Greeks, notably Democritus. It is unclear whether the Greeks took atoms terribly seriously or whether atoms were just a debating stance, but at any rate, nobody did anything useful with the notion of atoms for about 2,000 years. What Kepler did, using only thought experiments and folklore, was to trace the evident macroscopic regularities of snowflakes to some hypothetical microscopic regularity. He realized that if matter is made from tiny, identical particles, then its large-scale structure will depend on how those particles are arranged, not on individual shapes of those particles. As with life, the overall organization matters far more than the precise components.

Suppose, for simplicity, that the particles are small spheres. How do spheres naturally pack together? Spheres are three-dimensional, but snowflakes are rather flat: Their constituent particles—if there are such things—must lie close to a plane. Therefore, it is easiest to think about this question in two dimensions, using small discs—say, coins of the same value. Put one such coin down on a table (Figure 12a), and see how many you can pack around it. The answer is six, and they fit precisely (Figure 12b). Continuing in this way, you can pack coins together in a tight honeycomb pattern (Figure 12c). It is highly plausible that this pattern packs more coins into a given region than any other—it is the *densest* packing. Indeed this is not just plausible, but also true, although it was not proved in complete logical detail until the 1930s, by the Hungarian mathematician Fejes Tóth. From Kepler's point of view, the important feature is the sixfold symmetry of the packing. If ice is made from tiny units that pack together efficiently and lie on a common plane, then an inherent tendency toward sixfold symmetry is unavoidable. Principles of efficiency imply regular geometry.

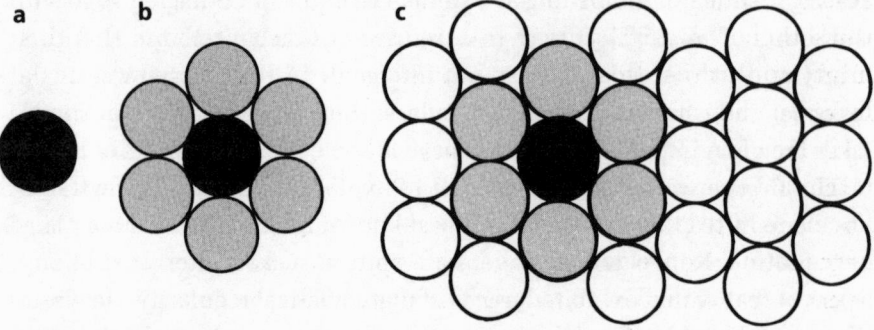

Figure 12
(a) How many coins pack around one?   (b) The answer: six.
(c) Next, extend the packing to a hexagonal lattice.

The development of crystallography has made it possible to check Kepler's contention. Matter *is* made of atoms, and in solids, they *do* pack together efficiently because nature prefers configurations that minimize energy. In an ideal world, the result of the efficient packing of atoms would be a regular mathematical structure known as a *lattice,* which is like three-dimensional patterned wallpaper, with the same unit repeating along three independent directions. In the real world, atomic structure is an approximate lattice, with small errors known as "dislocations." Crystal form is rich: Mathematicians have proved that there are precisely 230 fundamentally different lattices, with that many different types of symmetry. Now, a molecule of water contains two atoms of hydrogen and one of oxygen, so the crystal lattice of ice cannot be as simple as a honeycomb formed by copies of a single atom. However, it is very similar in form, as was discovered in 1878. The ice lattice turned out to be a three-dimensional arrangement in which each molecule of water is surrounded by four others, and—reflecting Kepler's insight—the oxygen atoms are arranged in layers, each layer forming a honeycomb. The hydrogen atoms complicate the picture but do not change its essence. In particular, the layers are planar, with sixfold symmetry—and this symmetry is one part of the explanation of the shape of a snowflake.

The molecular lattice does not explain the beautiful branching structure, however. That explanation is more subtle, to do with the way in which ice crystals grow.[2] It is in this explanation that we get our first hint of the intricacies of the structures provided by the mathematical laws of physics, without any further assistance or effort. Ice crystals grow by adding clusters of atoms to the outer surface of the lattice, and (as with the soap bubbles), they do this in a manner that minimizes energy. It might appear that this process should result in regular growth in flat layers, perhaps leading to a standard hexagon with straight sides (Figure 13a). However, John Day showed in 1962 that long straight sides do not minimize energy; instead, ripples develop (Figure 13b), a phenomenon known as "tip-splitting instability." This rippling changes the basic shape of the crystal from a flat-sided hexagon to a six-pointed star. This change is merely the first of many possible changes of shape. The same tip-splitting instability implies that growing arms of the star may in turn sprout new branches; those branches, when they reach a critical size, may also split; and so on. The result is known as "dendritic growth," and it is what leads to the fernlike appearance of most snowflakes.

We have now explained both the symmetry and the intricacy of snowflakes—but not their individuality. *Why* are no two snowflakes alike? The precise details of the tip-splitting instability depend on the surrounding environment, especially the atmospheric humidity. The more humid the air, the faster the crystal grows. Imagine a tiny hexagonal seed crystal inside a thundercloud, collecting new water molecules around its edge. Because the seed is so tiny, the conditions at all six corners are pretty much the same at any given instant, but because the snowflake is being blown every which way inside the thundercloud, those conditions change from one instant to the next. So all six corners

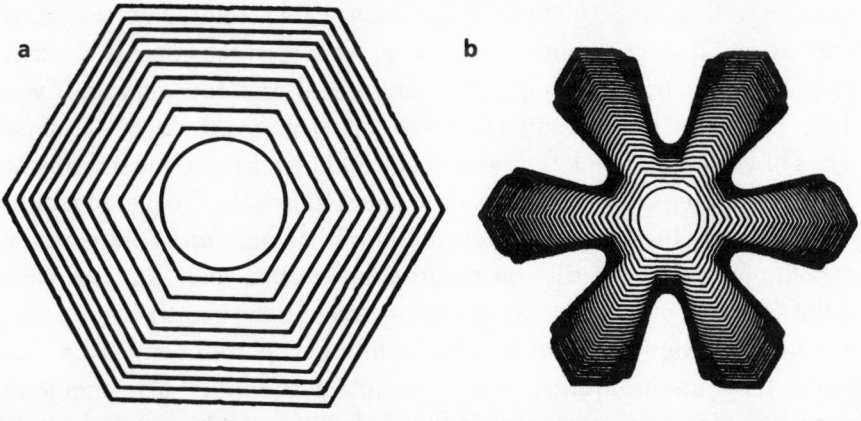

Figure 13
The hexagon (a) is not a minimal-energy configuration for an ice crystal; instead, ripples develop (b).

start to sprout, in identical patterns; then the conditions change and they all continue to grow in a slightly different pattern—but one that is still repeated at all six corners. So the hexagonal symmetry is maintained, but the shape itself varies according to the flake's precise history. The number of histories is pretty much infinite: end of story.

It's important to understand that the mathematical laws of crystal growth explain *all* of the intricacies of a snowflake; these intricacies are not just limited to the obviously mathematical hexagon with straight sides. Some of the intricacies, such as tip splitting, arise spontaneously without any obvious outside intervention; others are caused by external influences, such as variations in humidity inside a cloud, but even then, it is mathematics that controls the effects of those influences on the snowflake's form. The beauty of a snowflake is *all* present in its mathematical laws.

The form of a snowflake is primarily a *spatial* pattern; there is no interesting regularity that involves time. Another familiar and important pattern, however, is intimately bound up with regularities in time, as well as in space. This pattern is the wave, one of the commonest regularities in nature. In most parts of the world, if you stand on a beach and look at the ocean, you see not a flat calm, but row upon row of waves, rolling up to the beach and breaking in clouds of spray. Real ocean waves are very complicated, but many of their most important features can be derived from simple mathematical models.

Take a longish length of rope and tie it to a door handle; hold the other end and pull the rope fairly—but not completely—taut. Now jerk your hand up and down one time. You will see a wave, which appears to travel along the rope, bounce off the door, and travel back toward you. If you shake the end of the rope up and down repeatedly, you will see a whole series of waves, all much the same shape and traveling at much the same speed, following one after the other at equal intervals. The mathematical jargon for "having a regularly repeated form" is *periodic*. Forms can be periodic in space (e.g., tiles on a bathroom wall), in time (e.g., the cycle of the seasons), or in both (e.g., waves coming up the beach).

The traveling wave is one of the commonest patterns in the natural world. It occurs in liquids, as we have just described. It also occurs in gases: Sound is a traveling wave in the air. Light, radio, X rays, and microwaves are all traveling waves of electromagnetism. You can also find traveling waves in sand dunes, in the mud on the bottom of rivers, and even in cloud formations. In just two areas in the world you will also find forests in which dying trees form slow-moving wave patterns.[3] Those areas are around the Yatsugatake and Chichibu Mountains of central Japan, and around various Appalachian peaks running from New Hampshire to Canada. The waves take the form of gray crescents, advancing slowly but inexorably through the forest. Within each crescent, trees die and fall. Behind the crescent, vigorous saplings sprout. These waves are caused not by pollution, but by a naturally occurring fungus. The waves take many years to change their position, and dead trees are eventually replaced by new ones; nonetheless, the underlying mathematics is the same as for any other wave. Waves of dying trees, important for forest management, combine a biological process with a mathematical one.

Along a length of rope, there is really only one way to arrange periodic wave patterns: They must come one after the other. Only the shape and the size of the wave can be varied. On a surface such as an ocean or a sandy desert, however, there may be many different kinds of wave patterns. The simplest ones are formed by waves arranged in parallel lines, like the lines on a sheet of writing paper. When waves roll up a beach, they are usually arranged in roughly parallel lines. Some typical patterns of dunes[4] found in deserts are illustrated in Figure 14. Dunes are an important part of the desert environment. Creatures that live in deserts exploit—and are constrained by—the mathematics of dunes, from the moment those creatures begin to evolve.

Waves can do more than travel in straight lines. You already know that when you throw a stone into a pond, you get expanding concentric rings. Patterns of expanding rings also occur in a chemical experiment called the "Belousov-Zhabotinskii (or BZ) reaction,"[5] a type of chemistry that biologists now think is closely related to the formation of animal markings such as spots and stripes. This is not to say that animals often display concentric rings, despite the Far Side cartoon (by Gary Larson) in which one deer commiserates with another one, which bears an unfortunate target pattern on its hide: "Bummer of a birthmark, Hal."

In 1958, B. P. Belousov reported that certain mixtures of chemicals can undergo an *oscillatory reaction*—one that cycles repeatedly through the same sequence of changes. It would have astonished the chemical world—except that it was so revolutionary a discovery that the chemical world didn't want to listen, and the work never made it into print. At that time, chemists thought they knew that such things could never happen because they appeared to produce order (pattern) from disorder (no pattern). The apparently spontaneous generation of order from

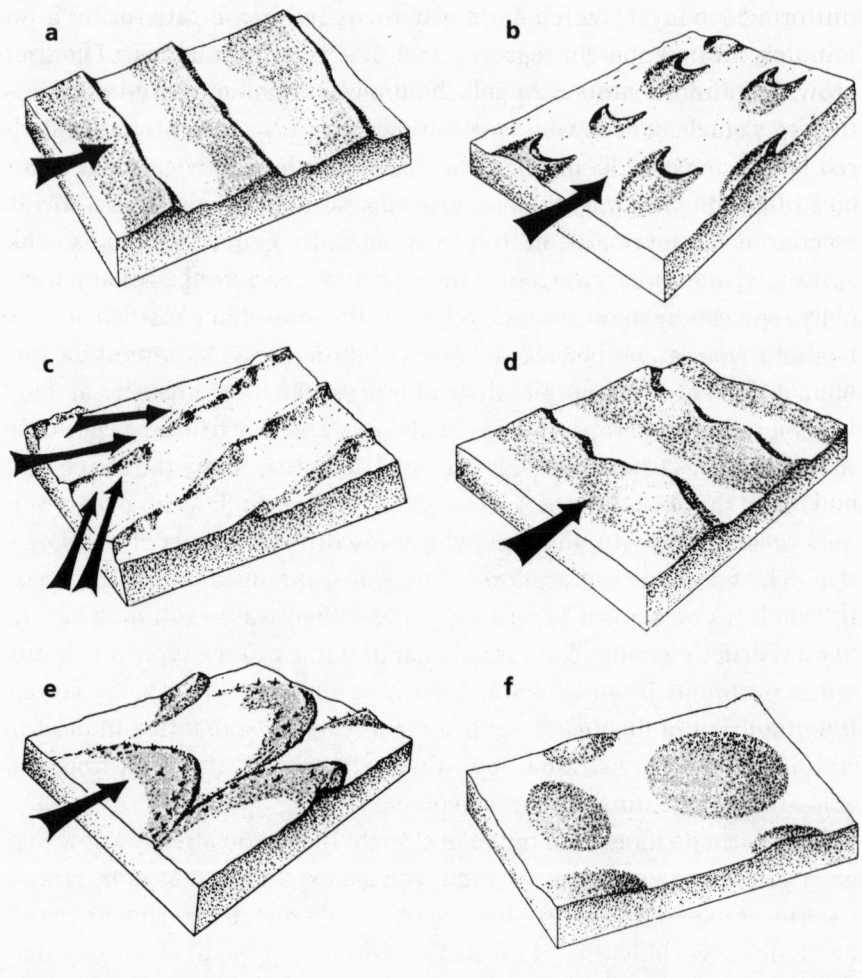

Figure 14
A selection of sand-dune patterns: (a) transverse dunes; (b) barchan dunes; (c) linear dunes or *seifs;* (d) barchanoid ridges; (e) parabolic dunes; and (f) dome dunes.

disorder was thought to be impossible. By 1963, however, A. M. Zhabotinskii had improved the choice of chemicals to make the effect more obvious and repeatable, and he showed that Belousov was right all along.

The experiment has since become a classic. As refined by Art Winfree and Jack Cohen, to improve its reliability, the experiment requires a mixture of four chemicals in fairly precise proportions.[6] If these chemicals are mixed together and placed in a shallow dish, at first, they form a uniform blue layer, which suddenly turns reddish brown. After a few minutes, for no apparent reason, a few tiny blue spots appear. The spots grow, and their centers turn red. As the blue rings move outward, and the red center spot expands with them, new blue spots emerge at the red spots' centers. Soon, the dish is filled with concentric rings of red and blue, all slowly growing and colliding with each other. The BZ reaction is more versatile still: As well as making rings, it can also make spirals. The simplest way to create spirals is to move the tip of a metal object, such as a needle, across the rings, to break them. You can also use heated wires or laser beams, or you can just tilt the dish slightly for a moment. Now the patterns begin to coil up around the broken ends, leading to a number of slowly rotating spirals, which interact with each other in complicated ways (Figure 15). Mathematicians can explain the rings and spirals using simple but subtle reaction-diffusion equations.[7]

These and a million other patterns speak volumes for the ability of the inorganic world to generate elaborate and beautiful forms. Where does this ability come from? In large part, it comes from a simple but fundamental principle, which governs the deep structure of the physical universe: symmetry. It was Albert Einstein, above all, who emphasised the importance of symmetry in our universe. He argued that truly fundamen-

Figure 15
Spiral patterns in the BZ reaction.

tal laws of nature must be the same at all times and in all places: That is, the laws must be perfectly symmetric. From this principle, he deduced, among other things, his theory of relativity. Modern physics has confirmed that at its deepest levels, the universe runs on symmetric lines. Principles of symmetry govern the four forces of nature (gravity, electromagnetism, and the strong and weak nuclear forces that act between fundamental particles); the quantum mechanics of elementary particles; the nature of space, time, matter, and radiation; and the form, origin, and ultimate destiny of the universe. We don't know *why*, but we are pretty sure that it is so.

Symmetry offers a simple explanation of regular patterns, such as crystal lattices, because such patterns are themselves highly symmetric. We find it easy to believe that the symmetry of the pattern reflects the symmetry of the laws that lie behind it. In addition, however, we are now beginning to appreciate that the fundamental symmetry of the universe can also offer an explanation of less regular patterns, too.

The essence of the matter is to ask how a universe based on perfectly symmetric laws can develop a diversity of different structures in space and time. Surely, if the laws are the same at all points of space and time, shouldn't that imply that the *universe itself* must be the same at all points of space and time? How can different parts of the universe behave in different ways if they are all following the same basic laws, starting from the same point—the Big Bang? How can intricate forms arise *spontaneously*—that is, without any of their complexity being generated in advance of their emergence—no complicated recipes, no complicated geometry, no complicated initial conditions . . . ?

The explanation of the diversity of form in the universe relies on symmetry, but it requires one further ingredient: stability. A system's state is said to be stable if it persists even when the system is disturbed by small random influences. Correspondingly, it is unstable if small disturbances cause big changes. In the real world, there are always tiny disturbances— such as the molecular vibrations that we call "heat." Usually, anything that we can actually see happening is stable.

The standard example of instability is a needle balanced vertically on its tip on a hard surface. In principle, if you could get it poised in *just* the right position, it would remain balanced forever. In practice, whatever you do, it falls over. The reasons are twofold. First, if you make the slightest error in positioning the needle, then that error becomes amplified by the force of gravity, so the needle will fall. Second, even if you could get the needle into that elusive ideal position, then the slightest breath of wind, or simply the vibration of the needle's own molecules due to heat, would disturb it enough to make it fall over. In contrast, a needle lying on its side is stable: if you leave it alone, it doesn't move, and if you give it a small push, it stops moving almost as soon as you stop pushing it.

Over the past few centuries, mathematicians have slowly come to recognize that instabilities in *symmetric* systems often give rise to patterns. Instabilities cause the overall symmetry of the system to break, meaning that what follows the instabilities is less symmetric than the overall system. The tip-splitting instability of ice crystals, which leads to the fernlike dendritic patterns of snowflakes, is a breakage of symmetry: Instead of forming a straight line, the boundary of the crystal develops wiggles. For a more homely example, think of a weight attached to the top of a vertical spring. This system has left-right symmetry: If the spring bends to the left, it exerts exactly the same force as if bent to the right. Also, if the spring is strong, compared with the size of the weight, then the *state* of the system will also be symmetric: dead vertical, straight up the symmetry axis. However, if the spring is weak, compared with the size of the weight, the vertical, symmetric position becomes unstable. Now the system seeks some other state, such as flopping over toward the attached weight. This state is no longer symmetric: It is biased in the weightward direction.

Where has the symmetry gone in this case? Well, for every state that flops to the left, there's another one just like it that flops to the right. The symmetry has been spread across several different states, instead of being concentrated in one. Usually, however, we see just one of those states at a time, so the overall symmetry is hidden from view.

Let's consider a different example. The surface of a duneless desert is flat and featureless—a highly symmetric state in which every position is exactly the same as any other. When that symmetry breaks—and it takes little more than a breath of wind to achieve this—the symmetric state becomes unstable. A little bit of sand piles up here, a shallow hole appears there. These changes to the surface affect the flow of air, and the disturbances are reinforced. Soon, huge dunes build up. However, because the original system, the hypothetical flat desert, is highly symmetric, some of that symmetry remains in the dunes. That's what gives them their striking patterns.

The breaking of symmetry lies at the heart of most of our understanding of pattern formation.[8] Mathematicians only formalized their understanding of symmetry in about 1830 or so. At the same time, they realized that a symmetry is not a thing, but a transformation—a motion, a deformation, a way to move constituent parts around. A symmetry is not just any transformation: It is a way to transform an object so that after it has been transformed, it looks exactly the same as it did to begin with. For example, the symmetries of a square include rotating it through a right angle: If you do that, you can't tell the differences among the rotated forms. This means that the transformation "rotate 90°" is a symmetry of the square. In contrast, if you rotate the square through 45°, you get a diamond, so that's not a symmetry.

What about something else—a circle, say? A circle has two kinds of symmetry: rotation about its center through any angle, and reflection on any diameter (axis). Each shape has its own set of symmetries; some shapes have very few symmetries, but some have a lot. The more symmetries a shape has, the more regular its pattern seems.

There are many kinds of symmetry transformation, including four main types (see Figure 16): *Translations* move objects bodily sideways without rotating them. *Rotations* keep one point (the center) fixed and move everything else through a constant angle, relative to that center. *Reflections* have the same effect as looking at the object in a mirror: The

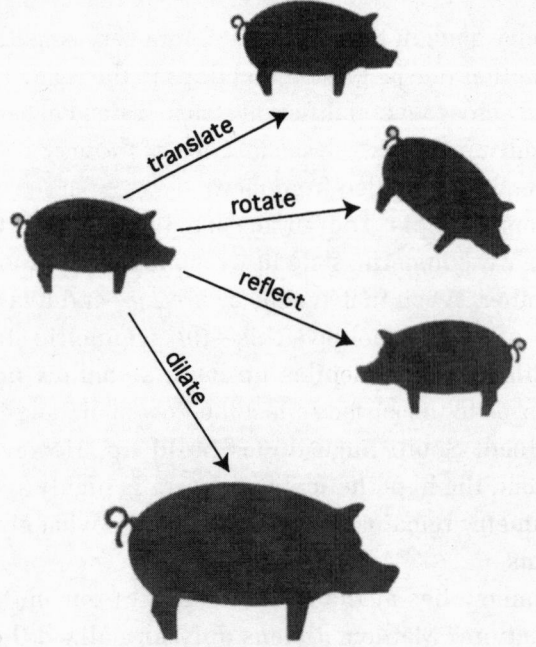

Figure 16
The four main types of symmetry transformation: (a) translation, (b) rotation, (c) reflection, and (d) dilation.

position of the notional mirror is known as the "axis." *Dilations* again fix a point—but now everything is expanded or shrunk, relative to that point, by some constant scale factor, in the same way that a map is a shrunken version of the territory that it depicts. Dilations are of particular significance in biology because they capture the effect of changes of scale.

Symmetries are important because they constrain the behavior of the physical universe. Broken symmetries are important because they explain unexpected changes of form—such as the tip splitting of ice crystals and the spinning spirals of the BZ reaction. Many important changes of form in biology arise through the breaking of symmetry; in such cases, mathematics offers clear and simple guidance on what we should expect, whereas the biological mechanisms are often distinctly puzzling. For instance, when a frog embryo develops, it undergoes a dramatic change in shape called "gastrulation"—stomach formation (Figure 17). Just before gastrulation, the embryo has developed to the *blastula stage,* in which it is a spherical mass of about a thousand tiny cells. At gastrulation, however, strange things begin to happen, for no obvious reason. A circular dent appears in the surface, beginning as a smile ∪, and then that region caves in, folding under and stretching to create a double-walled shape like a wide-bodied pot with a narrow spout.

The biology of gastrulation is complex, and it gives the impression that the blastula could, in principle, change into any shape whatsoever, depending on the genetic instructions that it is given. The mathematics of gastrulation, however, suggests that the observed changes in shape pick up on a ready-made inorganic pattern, a kind of buckling of the blastula under physical forces. In general terms, what must be happening is that the symmetry of the blastula breaks from a spherical to a circular pattern. Mathematical calculations show that exactly the same thing happens if any kind of spherical shell, a Ping-Pong ball, say, is compressed by uniform forces directed toward its center (Figure 18).

Yes, yes, I know: A frog embryo is more complicated than a Ping-Pong ball, and genetics and many other influences are busily at work, making sure that the buckling happens in just the right manner.[9] Equally clearly, frog development is exploiting a type of behavior that physics presented to biology, free of charge, hundreds of millions of years ago. The laws of physics, which have been with us since the dawn of time, imply that a spherical shell is a natural stable configuration for a collection of small particles, weakly attracted to each other. They also imply that—give or

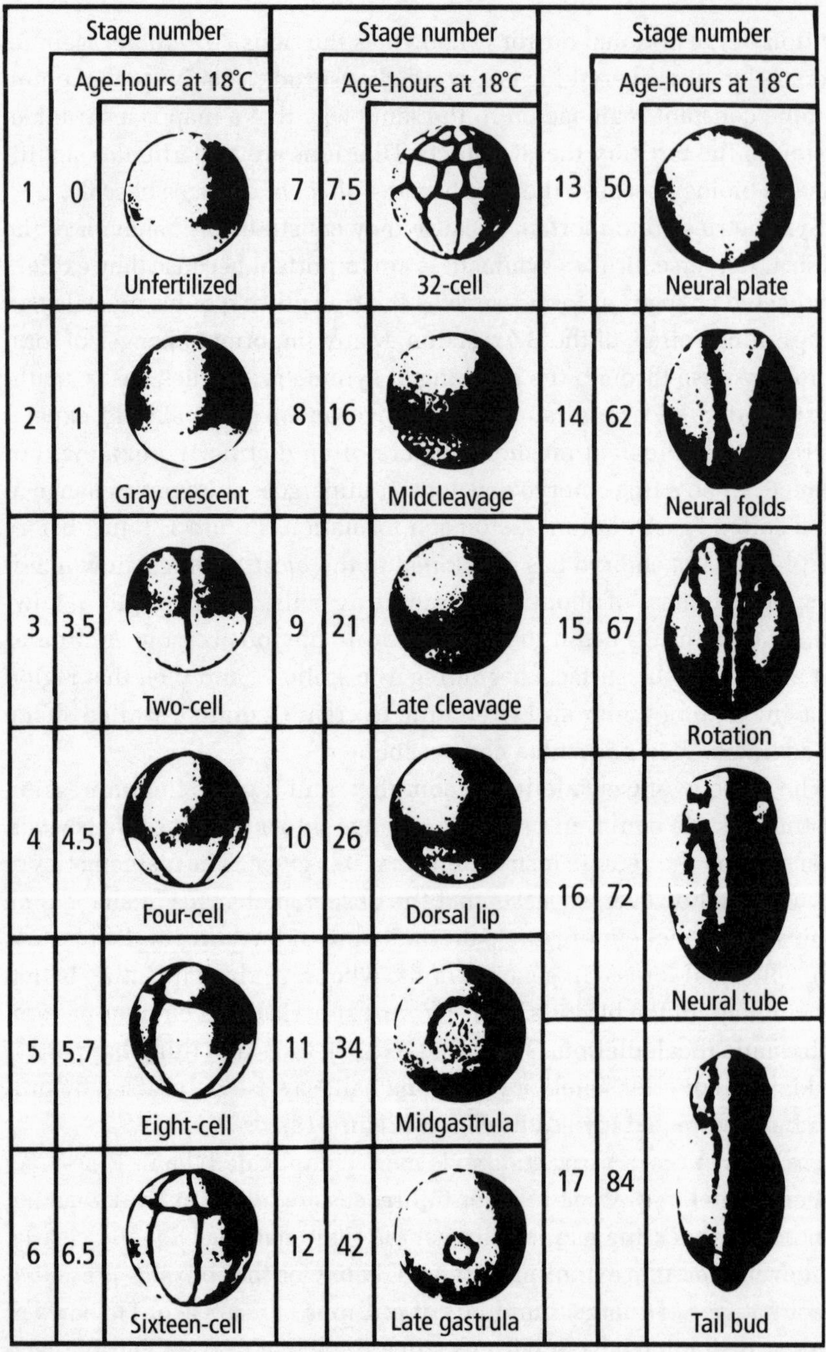

Stage number			Stage number			Stage number		
Age-hours at 18°C			Age-hours at 18°C			Age-hours at 18°C		
1	0	Unfertilized	7	7.5	32-cell	13	50	Neural plate
2	1	Gray crescent	8	16	Midcleavage	14	62	Neural folds
3	3.5	Two-cell	9	21	Late cleavage	15	67	Rotation
4	4.5	Four-cell	10	26	Dorsal lip	16	72	Neural tube
5	5.7	Eight-cell	11	34	Midgastrula	17	84	Tail bud
6	6.5	Sixteen-cell	12	42	Late gastrula			

Figure 17

The onset of gastrulation in a frog embryo (age 34 hours).

Figure 18
When a sphere
first buckles, it retains
circular symmetry
about some axis.

take a few technicalities—whenever such a spherical surface buckles, it does so by growing a circular indentation and folding inside itself.

It is entirely reasonable, therefore, for a prototype frog embryo to be spherical. On the other hand, it can't *stay* spherical, or else you'd end up with a spherical frog. So the symmetry has to break. Physics already provides a natural way for that to happen: Induce an instability in the shell, and allow it to buckle *spontaneously* into a circular dent, collapsing in on itself. This scenario comes free of charge to frog biology. Frog genetics may trigger the instability, and if necessary, it can evolve to modify the precise buckled form, but it doesn't need to tell the blastula how to buckle. Similarly, the genes for an elephant do not have to tell it to obey the law of gravity if it falls off a cliff. This means that you cannot understand gastrulation in purely genetic terms. You *must* build in the mathematics of buckling, too, or, more precisely, you must build in whatever analogue of buckling applies to the material of a blastula.

I don't want you to think that a simple spherical surface is necessarily an adequate model of a blastula. Maybe the individual cells of a blastula influence the buckling. Maybe genetic instructions tell those cells to start migrating—that seems very likely. Nevertheless, a significant piece of what then happens comes from physics, not from genetics. That's all I'm trying to say here.

The breaking of symmetry is a universal principle, applying to all symmetric systems, and it tells us this: Systems that are physically distinct will behave in similar ways, their obvious differences notwithstanding, provided they have the same symmetries. Symmetry breaking runs very deep in nature. It explains not just patterns, but also analogies among patterns. And, as we will see, it explains and unifies a surprising variety of patterns in living creatures. Before we work our way up to whole organisms, however, we must begin with a closer look at their molecular basis, DNA. Here, too, we find that mathematical patterns are paramount.

3

The Frozen Accident

The mysteries of organic chemistry are great, and the differences between its processes or reactions as they are carried out in the organism and in the laboratory are many; the actions, catalytic and other, which go on in the living cell are of extraordinary complexity. But the contention that they are different in kind from ordinary chemical operations . . . would seem to be no longer tenable.

D'Arcy Thompson, *On Growth and Form*, Chapter IX

If there is life elsewhere in the universe, it may not have the same chemical basis as life on Earth. Indeed, it may not have any *chemical* basis at all: Just ask the plasma-vortex beings of Apellobetnees III. However, life on Earth does have a chemical basis, and a very specific one. Everything hinges on a remarkable molecule, DNA. It is DNA that enables living creatures to organize the processes and structures that physics automatically provides. Life won't work without the mathematical laws that govern physics, as I keep saying—but there has to be something more, and on this planet, it's DNA.

DNA chemistry involves a lot more than just one molecule. It's more like a complex of chemical factories. There is a whole support team of other molecules, which accept instructions from DNA, carry them out, even modify the DNA itself, often in drastic ways. It seems likely that the chemistry of life could not take place in anything much simpler than a cell: Indeed *viruses,* a kind of low-level protolife, cannot replicate their genetic material on their own. Instead, they invade a cell and hijack its chemical factory, reprogramming its factory so that the cell turns out more viruses.

DNA poses many problems for science. How does it make copies of itself, a step that is essential for life to reproduce? How does it represent genetic information? Is the *genetic code,* which converts DNA instructions into proteins, the only code possible—a frozen accident, as Crick put it—or is there some deep physical reason why that code is preferable to all others? How crucial is DNA to life, anyway?

The answers to many of these questions are intimately bound to mathematical patterns. Indeed, what could be *more* mathematical than the famous DNA double helix? Much of the mystery that surrounds DNA seems to depend on the deep geometry of molecules. Indeed, at the level of molecular biology, geometry calls the tunes. Understanding something as basic as the process by which DNA makes copies of itself turns out to require some very sophisticated—and very modern—mathematics. Some of this mathematics was only invented since the late 1980s; some of it still hasn't been invented.

I believe that if we could understand life well enough, we would see it as just one more page in the universe's catalog of patterns. Page one: circles; page two: the DNA double helix; page forty-two: life, terrestrial style; page forty-three: plasma-vortex life-forms; and so on. I can't take you that far, but we can get up to page three: viruses. Without doubt, viruses come very close indeed to being raw mathematical patterns, but even they need some genetics, too. This chapter is mostly about the geometry of DNA, and various associated mathematical questions. It ends by going one step beyond DNA, to the molecular geometry of viruses—page three of the pattern book—and it starts by asking a time-honored question: Where did life come from?

To keep the discussion simple, I'll focus on life as we know it: DNA-based life. Our answers to the question, and our willingness to accept other people's answers, are heavily conditioned by how we think about life. If we perceive a huge gulf between life and nonlife, it becomes very difficult indeed even to imagine that an answer exists. A stupendous act of special creation starts to look far more plausible than some straightforward series of entirely natural, though possibly slow and sometimes accidental, events in the normal physical world. The narrower the gulf between life and nonlife seems to be, however, the more ready we are to admit that there might be an entirely rational explanation, not only that life *could* develop from inorganic beginnings, but that it *did*.

We've established that inorganic processes can produce structures and behaviors that are much more complex, and more sophisticated, than we usually recognize or appreciate. The laws of physics provide some quite flexible building blocks, which life is free to use, and to modify, if only it can get itself going in the first place. This realization certainly reduces the gap between the living and the nonliving realms, but it does not bridge that gap. Life still seems enormously more complicated, more organized, more flexible, and more adaptable than nonlife. Of course, life has had about four billion years to reach this point in its development. Because life is not just highly complex, but seems also to have the ability spontaneously to become even more complex, we should not expect nature to have made the transition from nonlife to today's richly textured life in one easy step. On the contrary, we must expect it to have started from small beginnings.

Unfortunately, we can't go back and look at those beginnings, so we have to infer what might have happened from whatever evidence is still available. The closest we have to direct evidence is fossils. The earliest life-forms were too soft, and too tiny, to make good fossils. Many later organisms did survive, however, and we can even date them by studying the type of rock in which they are found, or by using complicated measurements based on radioactive substances trapped in the rock. What the fossil record shows is that over long periods of time, billions of years, more and more complex organisms have arisen. I don't mean that the simple ones go away and are replaced by more complex ones; the simple ones generally hang around, though they, too, may change, or die out. Every so often, however, creatures appear that are distinctly more complex than anything that came before them.

This tendency toward greater complexity complicates our task. We must explain not only how life started, but also why the fossil record shows that life has a general, though not universal, tendency to become more complex as the megayears pass. The standard explanation for the growth of complexity is Darwin's theory of evolution. His theory is the topic of Chapter 4, so in this one, we are free to concentrate on the fascinating problem of getting life started from inorganic materials and processes.

In fact, the origin of life no longer appears to be a particularly difficult problem. We know that—at least on this planet—the key ingredient is DNA. Life's basis is molecular. What we need is an understanding of complex molecules: how they might have arisen in the first place, and how they contribute to the rich tapestry of living forms and behavior. It turns out that the main scientific issue is *not* the absence of any plausible explanation for the origin of life—which used to be the case—but an embarrassment of riches. There are many plausible explanations; the difficulty is to choose among them. That surfeit causes problems for the question "How *did* life begin on Earth?" but not for the more basic issue, which is "*Can* life emerge from nonliving processes?"

Let's begin with one of the most puzzling aspects of life: its ability to reproduce. Rocks do not make more rocks—except by breaking into pieces—and planets do not make more planets. However, amoebas, daffodils, frogs, pine trees, and elephants make more amoebas, daffodils, frogs, pine trees, and elephants. We need to find out how they do it—not the details, but the essence. Without reproduction of some sort, life cannot possibly get going. Where does life's ability to reproduce come from, though?

There are precursors for reproduction in the laws of physics. A good example of a reproducing inorganic structure is the flame: Provided it is supplied with enough fuel (i.e., fire food), in the form of flammable material and oxygen, then a flame will spread, begetting new flames as sparks fly from burning materials to as yet unburnt ones. The flame even has quite a complicated structure, with an inner core, an outer halo of partially burnt material, and a rising plume of smoke. Nevertheless, we do not view flames as a form of life, and the reason is instructive: Although flames can reproduce, they have no heredity. A flame re-creates its form for itself, depending on its surroundings, as well as on the regular patterning inherent in physics, but it does not inherit specific characteristics from its parent. And *that* is much closer to what we expect of a living organism.

Nevertheless, the flame is a start. Not that it is likely that life began as a flame: The process may be along the right lines, but the materials are implausible.

DNA *alone* is just a molecule, a bit of chemistry, part of the inorganic world—one that can *replicate* (make *exact* copies of itself). Can we consider the first appearance of a molecule of DNA to be the origin of life? If so, we have to explain where such a curious and complex molecule might come from. Moreover, there is another difficulty: The manner in which DNA replicates, with its vast support team of other molecules, is so complicated that it seems unlikely that DNA was the first replicating system. It seems far more likely that DNA was a kind of chemical parasite that rode piggyback on some simpler replicating chemical system. As the simple chemical replicated, so did its attendant DNA.

One possible contender for the first replicating molecule is RNA—DNA's simpler cousin. The RNA-based theory first came to prominence in the 1980s, when Tom Cech and Sydney Altman discovered special RNA molecules now called *ribozymes.* These researchers theorized that once upon a time the Earth was an RNA world containing oceans of a kind of chemical soup. In these oceans, molecules collided, reacted, broke up, stuck, and grew. As time went on, quite complex molecules arose, among which was a rudimentary form of RNA. It so happened that this RNA could replicate. If it bumped into suitable smaller molecules, it could collect them, and in some manner or other it could stick these molecules together to make a copy of itself. Now here's the thing about replicators: They replicate. As soon as you've got one of them, then provided the appropriate raw materials exist to make more, it *will* make more. Lots more. Soon the ocean was full of replicating RNA.

All very well, but how come self-replicating molecules are possible at all? Isn't that the more important question? You can't just wave your hands and invoke such an unlikely kind of molecule as given. To deal with this difficulty, Stuart Kauffman has developed the concept of an "autocatalytic network" of self-sustaining chemical processes,[1] which suggests that while a single replicating molecule is rather unlikely, a collection of molecules that collectively replicate each other is almost unavoidable. The idea is that it's a rather amazing coincidence if molecule A makes copies of molecule A; but it's not at all unlikely for molecule A to make a different molecule B because that's what molecules do. Then B makes something, call it C, or maybe it makes both C and D. After a while, because there are only so many different molecules to go round, you start to find repetitions—maybe D makes A and E, say, and E makes B.

Now B makes D makes E makes B. So molecule B replicates—with a little help from its friends.

Kauffman showed that if you keep adding molecules like this, eventually it's very likely that some subcollection will close up, so that the molecules made by anything in that subcollection are also in the subcollection, and every molecule in the subcollection is made by *something* in the subcollection. At that point, the subcollection becomes self-sustaining and self-replicating—an autocatalytic network. As with the RNA-worldview, the same point applies to these networks: Once one such network has arisen, soon you find little else.

At this point, the distinction between replication and reproduction, which until now I've left implicit, becomes important. A system replicates if—in the presence of suitable raw materials—it can create a more or less *exact* copy of itself. Crystals can replicate: New layers of the atomic lattice grow on existing layers, and under suitable circumstances, these layers can be split off to form a new crystal with pretty much the same structure as the original. In contrast, a system reproduces if it makes *modified* copies of itself that can also reproduce in their turn. A central feature of life is that it reproduces, rather than merely replicating. If life only replicated, evolution could never have parlayed simple forms into more complex ones. Life reproduces because living creatures possess not just heredity, but *flexible* heredity.

Kauffman's autocatalytic networks replicate but don't reproduce. An alternative suggestion for an inorganic precursor to life, one that possesses a rudimentary form of heredity, was made by Graham Cairns-Smith: His suggestion is clay.[2] We normally think of clay as some kind of sticky substance that can be molded into any desired shape, but what we now must appreciate is that clay—like snowflakes, like unaided chemistry, like the rest of the inorganic world—is much more interesting than we usually give it credit for. Down on the molecular level, and on scales not much larger than the molecular, clays have an amazingly varied and adaptable structure. They can form winding helices, strange shell-like shapes, and stacks of flat platelets like pancakes in a fast-food outlet.

In the right circumstances, stacks of microscopic clay pancakes can also replicate, thanks to the laws of physics. (I know I said that mere replication isn't interesting for life, but wait.) Think of a stack of clay pancakes forming underwater, perhaps on a lakebed, where the currents are gentle and new material is constantly being deposited. The top pancake in the stack acts as a template for the formation of a new pancake one layer higher; this in turn acts as a template for the next layer, and so on. Eventually the whole pile gets so high that it can easily be toppled by a small disturbance. So part of it falls off and starts a new pile: This is replication—but, so far, without heredity.

However, suppose that the growing stack gets the pattern wrong, maybe because there is too little extra material to make the next layer properly, or maybe because a bit gets broken off. Whatever the cause, the top pancake changes its shape. Because it acts as a template for the next generation, that shape change is then faithfully transmitted to its descendants (Figure 19). When a few layers fall off, to start their own line of descendants, that changed form is *still* faithfully replicated.

I said earlier that crystals replicate, but not in a very interesting way: They have no heredity. Now I'm going to have to change my mind. Because the clay pancakes are in effect a type of crystal, we see that crystals can be equipped with heredity when they inhabit a more complex environment than a test tube. I'm not claiming that clay pancakes are life, although if I wanted to, I could start to make a case: For example, they compete with each other by dissolving and recrystallizing, so in principle, they could evolve. It is possible to imagine a world in which clay structures evolve into more and more complex forms, developing the ability to move, to engulf other clay structures—perhaps even to combine their heredity with that of other clays to produce what might be called "offspring," which differ radically from any of their parents. . . . Does such a world exist? I have no idea. I doubt it—but it's a big universe, and maybe somewhere a dozen galaxies away. . . It certainly raises the ante on the question "What is life?" It also suggests that DNA may not be a *necessary* part of the answer.

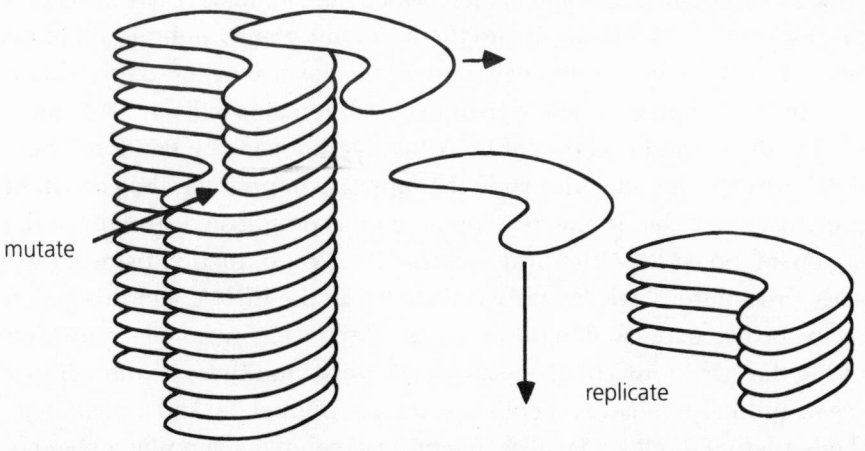

Figure 19
Replication and reproduction in a stack of clay platelets.

Be that as it may, clay pancakes definitely reproduce. It turns out that DNA could easily have piggybacked on clay because the DNA molecules are just the right size and shape to fit into molecular grooves in clay. Then DNA's complicated support team of chemicals might not be needed, not at first: The clay could take over the necessary tasks, gripping the DNA tightly in its grooves, copying the DNA along with itself. Then, with DNA firmly established, the support team could slowly evolve and take over the role of the clay.

Equally, DNA may have piggybacked on some autocatalytic network—or on something totally different. Instead, maybe it didn't piggyback at all—but if so, an extraordinarily complicated scenario has to get going all at once, which seems unlikely. Maybe DNA piggybacked on RNA, and RNA piggybacked on clay: In 1996, Jim Ferris discovered that long RNA molecules can form spontaneously on the surface of montmorillonite, a special kind of clay. As I said, the problem of the origin of life is not an absence of theories, but an overabundance.

Let's move on. Once life has originated, what makes it *behave* like life? DNA has two important functions: replication and programming the growth of proteins. We think that these functions are essential to all forms of life; we know they are essential to life on this planet. The first function, already mentioned, is replication. DNA is often described as a *self*-replicating molecule, but this is rubbish: DNA is no more self-replicating than a document sitting on top of a photocopier is. It replicates only by virtue of a whole host of subsidiary molecules, which cut strands of DNA into short pieces, unwind them, duplicate them, rewind them, and join them up again. It took molecular biologists about thirty years to sort out the main features of this process, and a great deal still remains to be understood.

It is the second function of DNA that really opens up Pandora's box. DNA provides a molecular code that appears to program the growth of proteins, a key step in the development of a complicated structure—the organism. So earthly life-forms manage their reproduction using a two-stage procedure: Their heredity is encoded in their DNA, and this coded recipe is then expressed in the actual form of the organism. The only kind of life that we know about—perhaps the only kind there is, though that seems unlikely—possesses both a *genotype* (DNA code) and a *phenotype* (body plan). For clay pancakes, in contrast, genotype and phenotype are the same thing: shape.

DNA's ability to replicate (reproduce) is based firmly on mathematics; it is a consequence of the molecule's geometry. As I mentioned earlier—

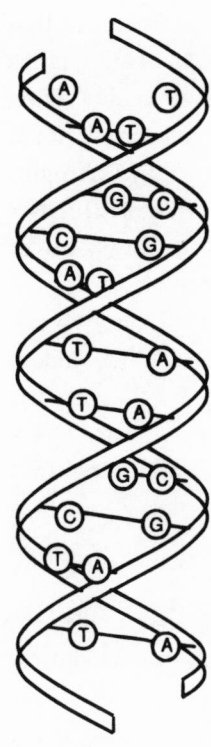

Figure 20
Cross-linking of
complementary
bases in DNA.

and no doubt you knew anyway—DNA is a double helix,[3] containing component submolecules that encode genetic information. These submolecules are called "bases," and there are four of them: adenine, thymine, guanine, and cytosine. Other components of DNA are deoxyribose (a sugar) and simple phosphorus compounds known as phosphate groups. The overall structure is long and thin, consisting of two strands twisted together in a double helix; each helical strand is made from the sugar and the phosphate groups.

Strands at the same level are cross-linked by pairs of bases, joined together by hydrogen bonds, shown schematically in Figure 20. The four bases have different shapes and sizes, and it turns out that they can fit together in the space available only when adenine is paired with thymine and guanine with cytosine (Figure 21). If we use the initials A T G C for the four compounds, then A always pairs with T, and C always pairs with G. Each strand can use all four bases, and on any given strand, the base pairs could occur in any order; but if one strand has base A somewhere, then the other must have T in the corresponding position, and so on. For instance, if the first few bases on one strand are A A C G T T T C G A T, then the corresponding bases on the other strand must be T T G C A A A G C T A.

The bases A, T, G, C are like the letters of a very limited, but highly important, alphabet. In the DNA of living organisms, these letters are assembled into long sentences, paragraphs, chapters. . . . A standard metaphor is "The Book of Life," and in those terms, the sequence of bases can be seen as a kind of molecular blueprint, a coded message written with the letters A, T, G, C, which specifies how the organism should be built, and from what ingredients. The metaphor has its faults, but it does capture the flexibility of DNA: You can write virtually any description you want into the DNA code book. Equally clearly, it explains why the DNA molecule must be huge: Animals are very complicated things. The DNA in each nucleus of a human cell is the equivalent of a kilometer (about 0.6 mile) of wire in an aspirin. These cramped quarters introduce

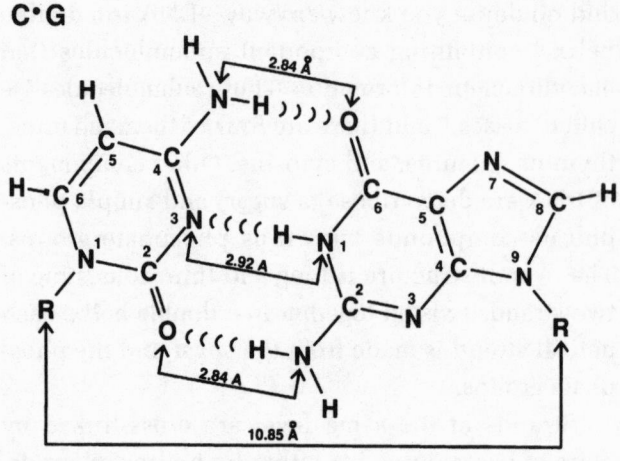

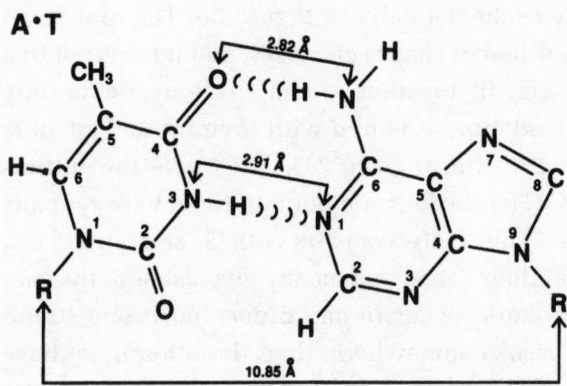

Figure 21
Watson-Crick pairing
of DNA bases.

a significant complication into the picture, by the way: The intertwined strands are not stretched out straight like a rope; they form a complicated tangled mass. Real DNA does not form perfect, straight helices—they are twisted, supercoiled, and pulled out into hairpin shapes and cruciform structures.[4] Indeed, my whole description is highly simplified: In addition to the standard links, there are nonstandard connections between base pairs, known by such names as "reversed Watson-Crick" and "Hoogsteen," and bizarre cross-linkages of deviant base pairs, such as "wobble," in which G links to T, not C. There are even triple- and quadruple-stranded forms of DNA. Those, however, are side issues, as far as the main story goes.

So far, we've started in the inorganic world and worked our way as far as a replicating molecule, DNA, which can act as a blueprint for an organism. However, we haven't yet worked our way up to an organism, as such. One way to bridge the gap between DNA and an organism is to introduce the cell. If we can build cells and get them to divide, then we can grow an organism from a single cell and make it as complicated as we like. It turns out that the structure of DNA has implications for cell division.

Because a cell incorporates a coded description of an organism, then whenever a cell divides, that description must somehow be copied into both of the resulting cells. The fixed pairing of A with T and G with C suggests an apparently simple mechanism: The double helix unzips into its two component strands. Then one strand begins AACGTTTCGAT, say, and the other has the complementary sequence TTGCAAAGCTA. Now there is room for new molecules of A, T, G, and C to attach themselves to the ends where the hydrogen bonds have been broken (Figure 22). Because of the matching rules, the *only* sequence that can attach to the free ends of strand AACGTTTCGAT is the complementary sequence TTGCAAAGCTA—not the old one, but a new copy, assembled letter by letter. By the same token, the only thing that can attach to the free ends of the existing strand TTGCAAAGCTA is a new complementary sequence AACGTTTCGAT. We end up with *two perfect copies* of the original double helix, each containing one strand from the original DNA double spiral.

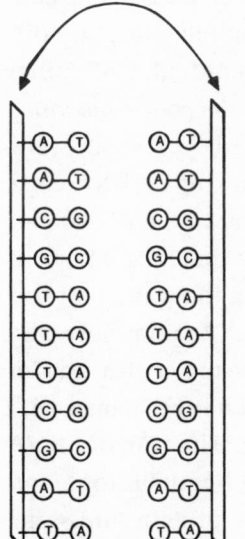

I said that in addition to replicating, DNA acts as a code for proteins. Let's see how. Protein production is probably the best understood aspect of DNA's repertoire of molecular computation, and it hinges on the *genetic code*. This code is the system by which DNA bases determine amino acids in proteins. The cell's molecular machinery reads the genome base by base, grouping the bases by threes, into triplets. So the sequence

$$\ldots AACGTTTCGATC \ldots$$

is split up as

$$\ldots AAC \bullet GTT \bullet TCG \bullet ATC \ldots$$

and so on. Notice that, in order to do this, the cell's machinery must know where to start and in which direction to proceed. A lot is known about how this is done, but for our purposes, it can be taken as

Figure 22

Replication of DNA (schematic).

given. There are four choices of base for each letter in the triplet, making a total of $4 \times 4 \times 4 = 64$ possibilities. The genetic code assigns each triplet to an amino acid—except that three triplets, ATT, ATC, and ACT, are assigned to the instruction STOP. Because there are only 20 amino acids, it is clear that the code is *redundant*—several different triplets may code for the same amino acid. For example CAA, CAG, CAT, and CAC all specify the amino acid valine, whereas GTT and GTC both designate glutamine. A given amino acid may be represented by between one and six triplets, and the genetic code displays an intriguing mixture of irregularities and patterns.

In order to discuss what these patterns are and how they may have arisen, it helps to know how the cell turns DNA base sequences into proteins. Proteins are made in cellular organelles called "ribosomes," *not* in the nucleus where the DNA recipe lives. Therefore, the instructions have to be transported over from the nucleus to the ribosomes. It would be possible but not very sensible to transport the nucleus's master copy (i.e., the DNA): Part of it could be hijacked en route, say by a virus. So the head office (nucleus) prudently retains the master copy and couriers just a photocopy to the protein factory (ribosomes). This copy is written using the other type of nucleic acid, RNA, which also has a sequence of bases of four types. The RNA bases are *almost* the same as those for DNA, except that one DNA base, thymine, is replaced in RNA by uracil, a slightly different chemical. It's like writing one letter in a different font, CAT rather than CAT; it affects the chemical mechanisms, but not the code's *meaning*. (In practice, by the way, uracil is denoted by the symbol "U.") At any rate, the cell copies the DNA sequence in the nucleus, making an RNA copy (known as messenger RNA), which the cell sends over to the ribosomes. Thousands of short photocopies are sent, rather than one long one, so if any copy gets lost in the mail, there's another copy on the way.

The protein-assembling trick depends on a bunch of relatively tiny molecules called "transfer RNA." There are 64 of these molecules, one for each triplet, and they all live at the protein factory. Each is preloaded with the appropriate amino acid for the relevant triplet. The transfer RNA molecule then waits until the corresponding triplet bumps into it, and then it glues the amino acid in place on the growing protein molecule. (In the case of STOP triplets, it glues on nothing at all.) This may seem to be a complicated way to go about making proteins, but nothing more direct would be reliable enough to hold together for billions of years—it would be too easily subverted.

It is here that a deep mystery opens up. As far as we know, the transfer RNAs could just as easily have been put together so that they implemented an entirely different genetic code. The reason that CAT is interpreted as valine is that the transfer RNA for CAT, which grabs onto that triplet because it has the complementary sequence GTA at one end, has molecular equipment for attaching valine at the other end. It would not be hard to put together a transfer RNA for CAT that attached glutamine instead. In other words, the genetic code could have been quite different without this having any serious effect on the organisms that populate the planet—just as Morse code could have assigned different sequences of dots and dashes to the letters of the alphabet without this having any effect on World War II. It's the meaning of the message, not the mechanism of the code, that matters.

Nonetheless, most organisms use the *same* genetic code.[5] Why? In 1968, Francis Crick proposed his "frozen accident" theory: Because the game is replication and the prize is life, the first code off the blocks will win. Maybe Crick was wrong, though. Maybe there were good reasons for nature to pick the genetic code that it did; maybe if you reran evolution, it would pick the same one again. Several competing theories have argued that there is a hidden rationale to the genetic code, and a new twist was added in 1993 by José and Yvonne Hornos,[6] who attempted to explain the pattern of the genetic code in terms of symmetry breaking.

Think about Morse code. Samuel Morse could have chosen to assign the same letter (*S*, say) to *every* sequence of dots and dashes. This system would have provided a highly symmetric code, but—of course—a totally useless one. Symmetric in what sense? For symmetries of codes, the relevant transformations are not motions in space; they are operations that swap sequences of code symbols around. A symbol sequence possesses such a symmetry if its *meaning* is unchanged by the swap. Now if all code sequences have the *same* meaning, then you can swap the symbols in any way you wish without changing that meaning. This is the sense in which my totally useless modification of Morse code is highly symmetric. My code could be made more useful by breaking the symmetry—for example, by assigning *S* to any sequence of dots (so that ·, ··, ···, ····, and so on would all mean "S"), *O* to any sequence of dashes, and *A* to any sequence containing both dots and dashes. The resulting code would no longer be completely symmetric; for example, swapping all dots and dashes would turn the message SOS into OSO. The new code would retain some of the original symmetry, however; for instance, the message AAA would remain unchanged. We can imagine further losses of symmetry that would lead, step by step, to the code that is enshrined in cryptographic history, with ••• for S, – – – for O, and so on.

Now think about the genetic code. We have already observed a key feature: The genetic code is redundant. That is, different triplets often code for the same amino acid. There is no great regularity to this lack of uniqueness, but a definite degree of symmetry—albeit imperfect—is clearly visible in the genetic code. Often, just the first two bases in a triplet determine the corresponding amino acid. For example GA? is always leucine, CG? always arginine. In short, the code for these amino acids is symmetric under changes of the third base. If this symmetry were perfect, then the 64 triplets would break up into 16 quartet triplets, such as GAC, GAG, GAA, GAT, with each triplet of the quartet coding for the same amino acid (but a different amino acid for each quartet). However, there are more than 16 amino acids, so sometimes the third base matters. Indeed sometimes the *second* base matters. Either way, the symmetry of the arrangement into quartets is broken.

How? The precise pattern of redundancy in the genetic code is messy. Instead of a tidy 16 quartets, there are three sextets (amino acids corresponding to six different codes), five quartets, two trios, nine duos, and two singles. The Hornoses assume that these numbers are determined by a chain of broken symmetries. The mathematical machinery reveals just eight possible chains; exactly one comes close to the correct numerical pattern. A final breakage of symmetry is needed to reproduce the precise pattern of redundancy found in the genetic code. This final breakage is somewhat empirical, and it suggests that the code may be a frozen version of one that ideally would have had 27 amino acids. The first broken symmetry—where the 64 triplets would code for just 6 amino acids—may represent a primordial version of the genetic code, the first step in its evolution. If the Hornoses are right, the genetic code is no accident. Pretty much the same code would arise again if we reran the origins of life.

Earlier, I said that the copying of DNA strands is quite a complex business. In fact, there are geometric reasons why it can't be simple. If you pull the two DNA strands apart in order to attach complementary ones, the strands wind around each other and get tangled up. Therefore, something else has to happen—such as special molecules, known as "enzymes," cutting the strands apart and rejoining them later. This aspect of DNA is currently being studied using some surprisingly sophisticated and very new mathematics: topology. *Topology* is often called "rubber sheet geometry," because it studies those features of a geometric shape that are unchanged if the shape is stretched, compressed, bent, or twisted—but not cut or torn. It is what geometry would be like if you drew the diagrams on a sheet of rubber: no fixed angles, no parallels, no straight lines. If this description makes the topic sound frivolous, here is a more sober description: the geometry of continuous transformations. *Symmetry transformations* keep objects the same shape—and the same size, unless we're talking about dilations. *Continuous transformations* can distort the shape and change the size; the only constraint is that the object has to remain in one piece. Because continuity is one of the deep properties of nature, topology is a fundamental mathematical tool.

Topology deals with geometrical features such as holes, knots, links, and boundaries. I now briefly describe two typical uses: a classical approach to DNA supercoiling and a modern attack on the role of enzymes in DNA chemistry. Supercoiling is just one of the ways in which the textbook double-helix model of DNA fails to represent the true complexity of that molecule. To experience supercoiling for yourself, take an elastic band and loop it loosely *between* the thumb and forefinger of each hand, pulling your hands slightly apart to keep the band taut. Now rub your right thumb against your finger to rotate one section of the band several times. This introduces a number of clockwise turns along one section of the band, and an equal number of counterclockwise turns along the other (Figure 23a). Now, pressing thumbs against fingers to prevent the band from twisting, bring your hands closer together. The two stretches of elastic twine around themselves like the tangles you often find in a stretchable telephone cord (Figure 23b). When this kind of thing happens in a DNA molecule, it is known as *supercoiling* because the standard helical form is already coiled (around itself). A supercoil is a coiled coil, and it is merely one of the simpler ways for DNA topology to make life difficult for the biologist. Supercoiling particularly occurs in *plasmids,* closed loops of DNA purified from bacteria, and it causes the loops to tangle up when viewed in an electron microscope (Figure 24).

The topology here is governed by two simple quantities known as the "linking number" and the "writhing number." A *link* is one turn in an elastic band after being twisted but before it is allowed to supercoil, and the *linking number* is the number of links that occur in the molecule. A *writhe* is the kind of tangle that occurs when the molecule supercoils, and the *writhing number* is the number of such tangles. These two types of twists in DNA may seem very different, but topology tells us that they

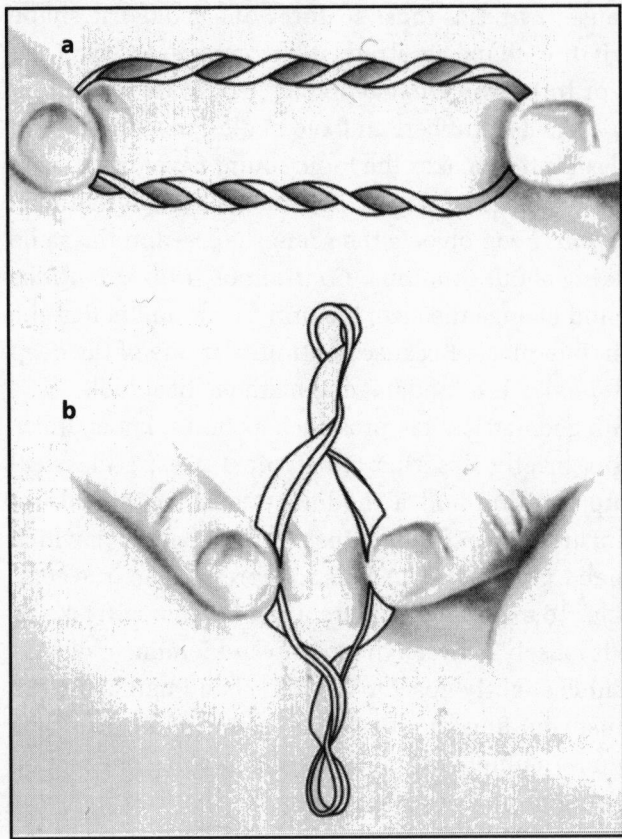

Figure 23
Supercoiling in an elastic band: (a) creating clockwise and counterclockwise links; (b) trading the links for writhes.

are actually closely related: In fact, the linking number is the sum of the writhing number and the number of helical turns in the DNA. This means that by making continuous deformations of the DNA, links can be traded for an equal number of writhes, or for extra turns in the helical structure itself. All three quantities can be computed rather simply by counting how the DNA strands cross each other, and the three together provide theorists with a powerful grip on the geometry of the DNA molecule.

The application to enzymes requires much deeper topological quantities, which capture not just the coarse overall geometry, but also the fine details of knotting and linking. Since 1928 or so, mathematicians have been startled to find how rich and difficult the classification and analysis of knots has proved to be. Apparently simple problems, such as distinguishing a reef knot from a granny knot, succumbed only to high-powered machinery and deep and complicated theories. Then, in 1984, a New Zealander named Vaughan Jones opened up a new chapter of

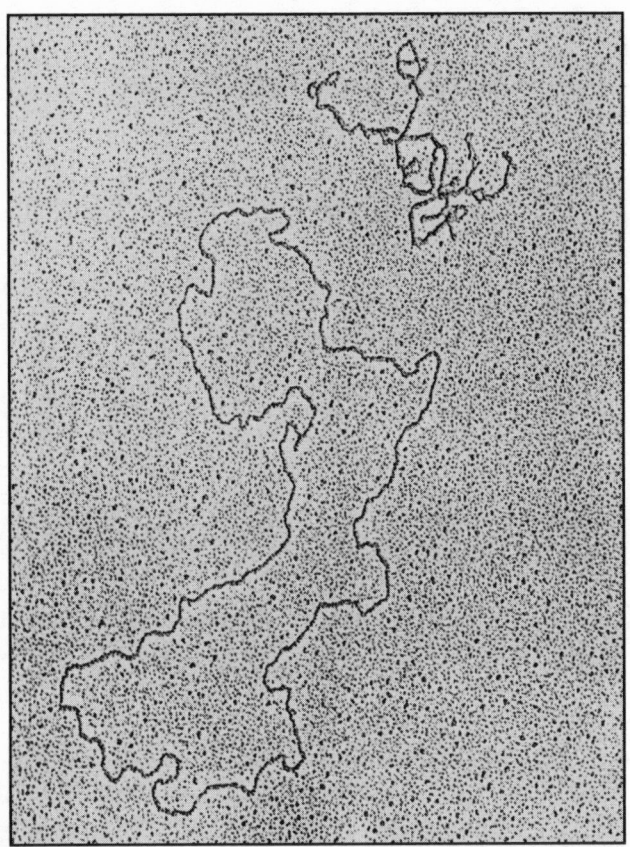

Figure 24
Electron micrograph of supercoiled DNA (top) and relaxed DNA plasmids (bottom). Each plasmid contains about 9,000 base pairs.

knot theory by inventing a totally new way to detect which knot is which—computable quantities that can distinguish many different knots from each other. These quantities, now called "Jones polynomials,"[7] were the tip of an iceberg, and generalizations and new discoveries are still flooding in.

Since the mid-1990s, this new machinery, together with some of the old machinery, has been harnessed to problems in DNA biochemistry, addressing the important problem of finding out the shape of strands of DNA *when in solution*. The traditional apparatus of X-ray diffraction requires the DNA to be in crystalline form, which doesn't help here; instead, biochemists inspect the actual molecule, squashed flat on a surface, and revealed in a high-powered electron microscope. As we know, the DNA double helix comprises two intertwined strands. When enzymes cut and then rejoin the molecule, the resulting strands form tangled knots and links. A fundamental problem is to work out which knot or link you are seeing; your interpretation of the experiment depends on doing so. For example, Figure 25 shows a single strand of DNA, which crosses itself 13 times. An awful lot of knots have 13 crossings—which knot is this one? In fact, it is a *torus knot,* meaning that it can be formed by winding a single strand over the surface of a doughnut, passing through the central hole as necessary, and never meeting itself until it joins up into a closed loop (Figure 26). The DNA molecule doesn't *look* like a neat, tidy torus knot because it has been deformed—the rubber sheet in action. Jones polynomials are not fooled by such deformations, however, and easily penetrate the knot's heavy disguise. With the aid of such topological techniques for distinguishing knots and links and for understanding how knots and links can change when they are cut and rejoined, biologists and mathematicians are beginning to unravel the secrets of DNA enzyme activity.[8]

Thus, even the replication of DNA is more complicated than the simple recipe, "pair off complementary bases," and the untangling of its geometry requires some of the newest and most sophisticated mathematics ever invented. Much the same goes for DNA's primary task of protein manufacture, which involves a lot more than just knowing the

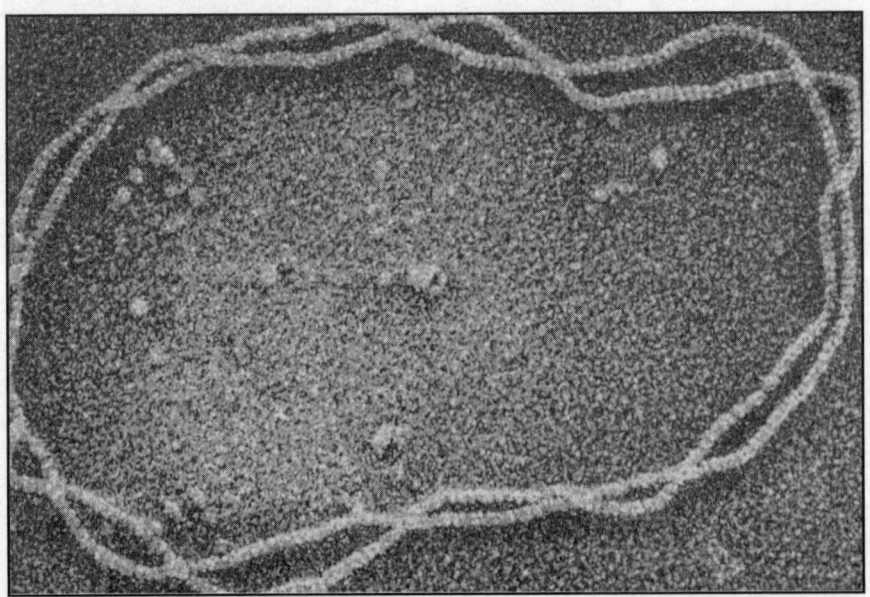

Figure 25
A DNA knot with 13 crossings.

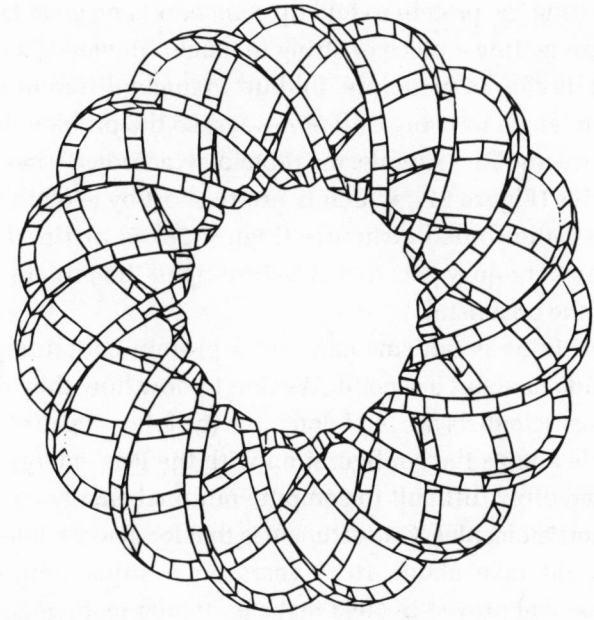

Figure 26
A torus knot that
winds 4 times
round the torus and
15 times through the
central hole.

genetic code. This time, not all of the necessary mathematics has been invented, although we know enough to see that the problem definitely is a mathematical one.

Specific segments of the genome—known as "genes," although the same word is used more broadly—encode recipes for proteins. A *protein* is a large molecule, say between a thousand and a million atoms, composed of amino acids. Exactly 20 different amino acids occur in living creatures (if we ignore a few very rare organisms). Like the bases of DNA, the amino acids in a protein are arranged in a specific sequence, but there is an extra wrinkle—quite literally. It is not the sequence, but the shape, that counts.

The function of most proteins is to manipulate molecules. For example, hemoglobin captures or releases atoms of oxygen; a key component in plant systems, chlorophyll, takes molecules of carbon dioxide and water and splits them apart, releasing some of the oxygen and incorporating the rest, along with carbon and hydrogen, into energy-rich compounds. These molecular manipulations depend on the protein's *shape* because they employ a kind of molecular lock-and-key principle: Incoming molecules such as oxygen or water fit into, or preferentially bind to, special nooks and crannies in the protein's surface. The shape of each protein is determined by how its amino acid chain folds up in three dimensions.

Getting the protein to fold up *somehow* is no great feat—no more difficult than getting a piece of string to tangle. However, a given chain of amino acids can, in principle, fold up in many different ways, just as a string can tangle in many different ways, so the problem is getting it to fold up *correctly*. For example, the protein cytochrome-c has a chain of 104 amino acids (Figure 27), which is pretty short by protein standards; nevertheless, the folded structure (Figure 28) is distinctly complicated—and unless biology gets that structure right, the protein won't work properly in the organism.

Living organisms can fold a protein containing a thousand amino acids in about a second. We don't know how they do it. We *think* (well, most scientists do, but I don't; see the following discussion) that the molecule adopts the configuration with the least energy.[9] Unfortunately, it is incredibly difficult to compute minimal energy configurations, even for short molecules. One estimate is that for cytochrome-c, such a calculation would take about 10^{127} years on a supercomputer. Indeed Aviezri Fraenkel proved in 1993 that a particular mathematical model of the protein-folding problem is what computer scientists call "NP-complete," which means "really really hard."

The difficulty here is that the number of potentially possible configurations is vast, and the minimal-energy configuration lurks among them

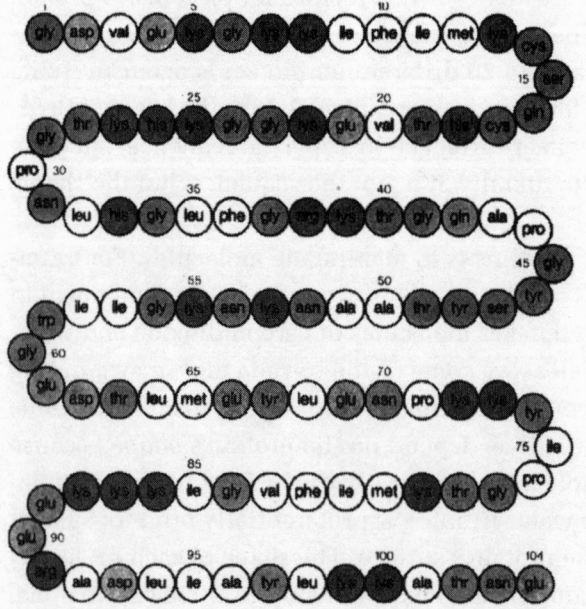

Figure 27
Linear amino acid sequence of the protein cytochrome-c.

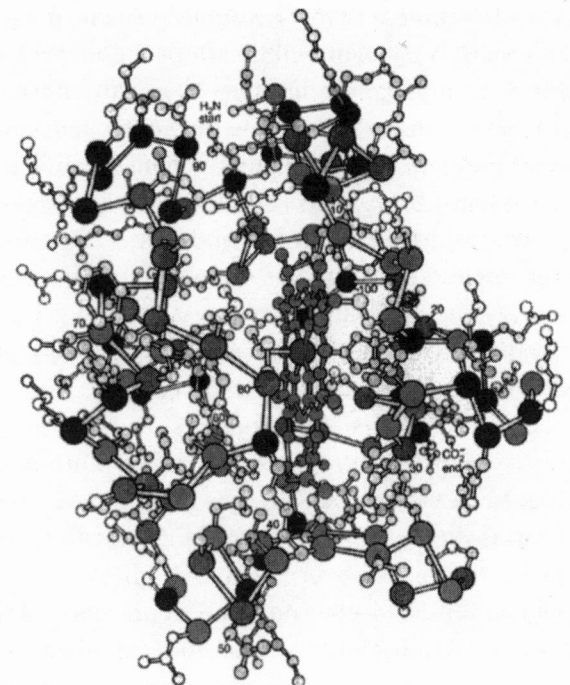

Figure 28
Folded structure
of cytochrome-c.

like a microscopic needle inside a haystack the size of a billion universes. Personally, I doubt that the biological processes solve that problem at all. I think that living organisms have found a quick-and-dirty method, which comes close to minimal energy—close enough to fool literal-minded human scientists into thinking that's what's going on—but doesn't necessarily get all the way there. Living organisms do not start with a complete linear chain of amino acids and then fold it up—which is what these horrendous computations try to do. Living organisms fold the chain of amino acids as they build it, one step at a time, and that must surely reduce the computational complexity. Also, I imagine, living organisms jiggle the partly formed protein every so often, to prevent odd protuberances from getting hooked up on extraneous loops. There are many NP-complete problems with approximate quick-and-dirty solutions, and I reckon we're looking at another one here. In fact, George Rose's new program LINUS employs *heuristic rules* (inspired scientific guesswork) to predict how proteins containing 1,000 amino acids will fold, and it does pretty well.

Once you've got a protein, what can you do with it? Virtually anything. You can carry oxygen to the lungs of an animal, or you can turn sunlight, water, and carbon dioxide into sugar in a plant. Another interesting thing that you can make with proteins is a virus. Viruses are a nuisance in everyday life because they can cause diseases, but they are godsends to biologists because they are a simple enough form of protolife that they can be investigated in considerable detail. One of the things that we learn from viruses is that it takes very little effort to put them together. Once the correct bits and pieces exist, the mathematics of intermolecular forces assembles them without further genetic intervention. Another thing we learn from viruses is that geometric constraints at the molecular level are very strong: Mathematics places stringent limits on the range of possible virus architectures.

The implications of mathematical constraints for structure are especially evident in *retroviruses* (viruses that contain RNA, rather than DNA). Most viruses use DNA as their genetic material, just as genuine life-forms do, but retroviruses use RNA, carrying it around in a protein case. Retroviruses hover in the twilight zone between life and nonlife. They can replicate—but only in the presence of the complex molecular machinery of a bacterium or another cell. They are molecular parasites, which hijack a cell's reproductive machinery for their own ends. They are also extremely mathematical in design: Common forms include flat discs, helices, and icosahedrons. To illustrate the possibilities, we consider two common viruses—*tobacco mosaic virus,* which infects the leaves of the tobacco plant, and *poliovirus,* which (at least until the development of effective vaccines) wreaked havoc on the human population.

Tobacco mosaic virus (Figure 29) is a hollow rod made from 2,130 identical protein units, with a coil of RNA some 6,400 bases long, running down the middle (Figure 30). In 1955, Heinz Fraenkel-Conrat and Robley Williams demonstrated that the tobacco mosaic virus's ability to function depends solely on its material constituents.[10] This helps bridge the gap between the inorganic and the organic worlds by showing that one of the most striking features of living material—replication—is here a consequence of molecular architecture; it does not require further intervention, either by genetics or by God. What Fraenkel-Conrat and Williams did was to separate the virus into its protein units and its RNA coil— reducing it to a collection of molecules that could, in principle, be synthesized from inorganic precursors—and then to show that when these components are mixed together in a test tube and left to their own devices, they *spontaneously* reassemble into a complete virus that can replicate inside a tobacco leaf cell. In short, they demonstrated that

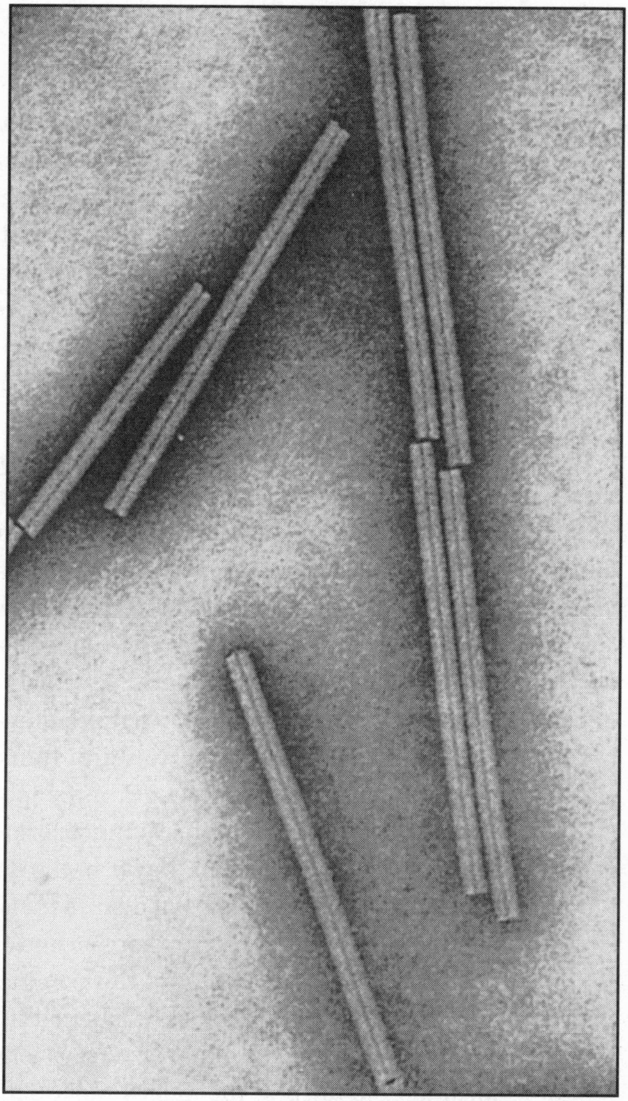

Figure 29
Electron micrograph
of tobacco
mosaic virus.

Figure 30
Structure of tobacco
mosaic virus: a helix
of identical protein
units surrounding
an RNA coil.

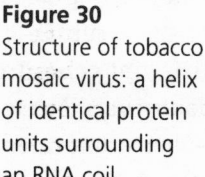

viral
RNA

protein
subunit

unaided chemistry can produce something as complex as the tobacco mosaic virus. A human chemist could set to work and *build* the tobacco mosaic virus from inorganic ingredients.

How does this spontaneous self-assembly take place? Virus growth, like that of crystals, occurs in two stages. First a suitable seed (germ) is constructed; then additional material is added to the seed, step by step. The obvious method for tobacco mosaic virus construction is to use the RNA coil as a scaffolding and to add protein units one by one, in a type of crystallization process. Curiously, however, nature does not follow the obvious method. Scientists discovered nature's detour because in the laboratory experiment, it takes much longer for the virus to reassemble by the crystallization route than it takes for the natural growth process. The main source of delay in constructing the tobacco mosaic virus is the time needed for creating the seed.

So how is a real tobacco mosaic virus built? It turns out that the protein units naturally assemble—as a result of symmetry breaking—into a variety of forms: isolated units, stacks of two or more units, discs, short stacks of discs, a cylindrical crystal, a dislocated disc (which looks like a lock washer), and a long helical stack. Precisely which form occurs depends on two ratios in the surrounding fluid: the salt concentration and the pH (acidity/alkalinity). The helical form (minus its central coil of RNA) occurs at low pH; however, the pH of the tobacco leaf cell is neutral (about 6.5). At neutral pH, helices do not form unless the RNA coil is also present. The most important player in the drama is not the helix, but the humble lock washer (dislocated disc), a transitional form between stacks and helices. If the pH is higher than 6.5 and the salt concentration is also moderately high, then discs are the dominant form. If the pH is then lowered, they dislocate into lock-washer structures. These molecular lock washers spontaneously come together to form a nicked helix with random gaps; then the turns of the nicked helix tighten up to remove the gaps and create a complete helix. This process, though it sounds more complicated, is a lot quicker than the laboratory method.

That protein assembly takes place at low pH, even without the RNA coil, is a natural inorganic pattern on which evolution can build. It does so in an amazing manner (Figure 31). The problem, for the virus, is to cause helical growth to occur at neutral pH, where it is *not* going to happen spontaneously without an RNA coil. How does nature use RNA to achieve this construction? The viral RNA molecule has a hairpin-shaped loop. This loop is inserted through the hole in the middle of a disc, and

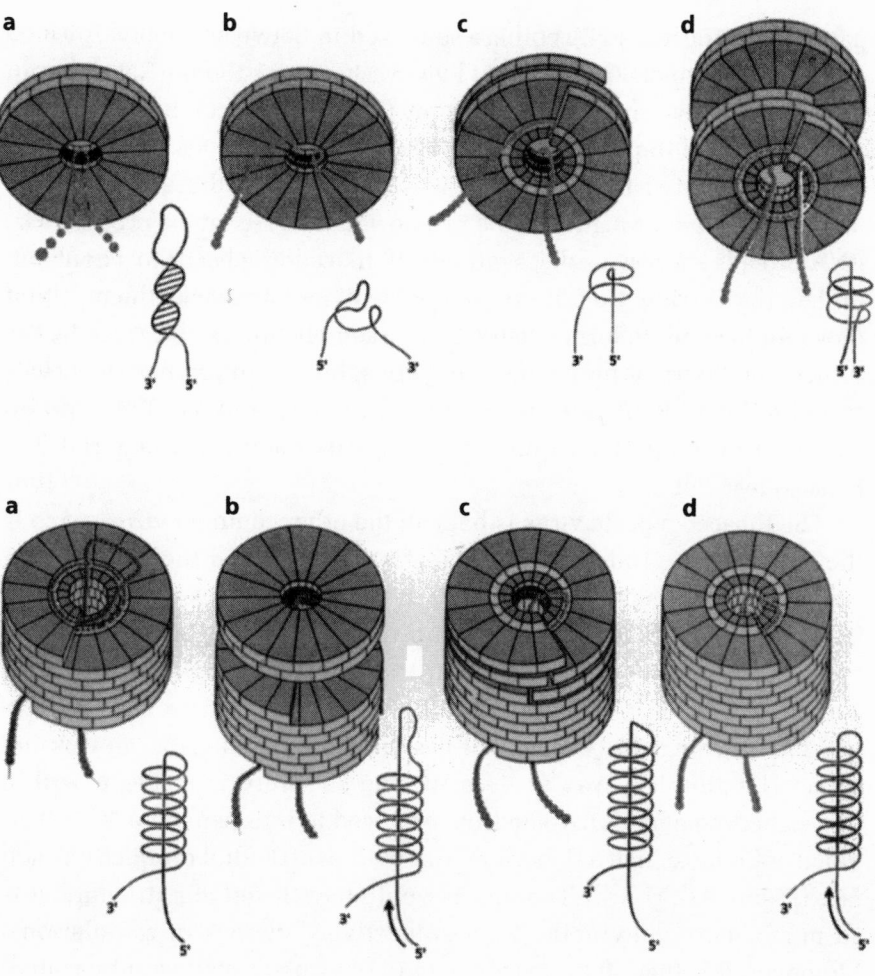

NUCLEATION of the tobacco-mosaic virus begins with the insertion of the hairpin loop formed by the initiation region of the viral RNA into the central hole of the protein disk (a). The loop intercalates between the two layers of subunits and binds around the first turn of the disk, opening up the base-paired stem as it does so (b). Some feature of the interaction causes the disk to dislocate into the helical lock-washer form (c). This structural transformation closes the jaws made by the rings of subunits, trapping the viral RNA inside (d). The lock-washer-RNA complex provides the start of the helix. Additional disks then add rapidly to the nucleating complex, so that the helix elongates to a minimum stable length.

Figure 31
How nature grows a tobacco mosaic virus.

parts of the trailing RNA coil are squeezed in between the layers of the disc, forcing it to dislocate into its lock-washer shape. So the RNA hairpin pulls through the central hole, dragging the rest of the coil behind it, with a few turns still trapped between the layers of the lock-washer structure. However, these twists of RNA can easily slide between those layers, so the hairpin can pass through the hole in a new disc, to repeat the process. The RNA threads successive discs onto itself, like adding beads to a doubled-up length of string; then it dislocates the discs and stacks them so that they join together to form a helix. This assembly process is driven by the lifeless geometry of molecular energetics, but evolution has selected a combination of RNA *plus* proteins because that produces the required structure in the preferred environment—the neutral pH material of a tobacco leaf cell.

The tobacco mosaic virus is helical; the other common virus shape is the icosahedron. Indeed, some scientists have called the icosahedron "nature's favorite shape" because it is so common in viruses—for instance in smallpox, polio, herpes, and turnip yellow mosaic viruses. The structure of poliovirus was discovered in 1986 by James Hogle, Marie Chow, and David Filman.[11] The virus is made from 60 copies of each of four protein units, arranged in a form that has the same symmetry as an icosahedron. The simplest way to describe the structure is to begin with a dodecahedron and an icosahedron, arranged to interpenetrate. The combined solid looks like a dodecahedron with pentahedral dimples on each face (Figure 32). D'Arcy Thompson would have loved this structure: It is so much more convincing than subjective drawings of radiolarians. Moreover, it is there for a *reason,* a kind of virus-crystallography principle: Just as minimum-energy configurations of large numbers of atoms form a crystal lattice, so minimum-energy configurations of a small number of identical units form a nearly spherical polyhedron. Of the regular solids, the icosahedron comes closest to being a sphere, but you can get closer still to a sphere by employing a mixture of five-sided and six-sided polygons. The modern soccer ball is an example: it is essentially icosahedral in shape, but it is a truncated icosahedron—one with the corners cut off. In such a polyhedron, there must always be precisely 12 five-sided faces: The number of six-sided faces is governed by a series of so-called magic numbers of a particular algebraic form; most numbers are not of this special form. The magic numbers less than 300 are 12, 32, 42, 72, 92, 122, 132, 162, 192, 212, 252, and 272. These numbers play a special role in the structure of a virus—just as Fibonacci numbers play

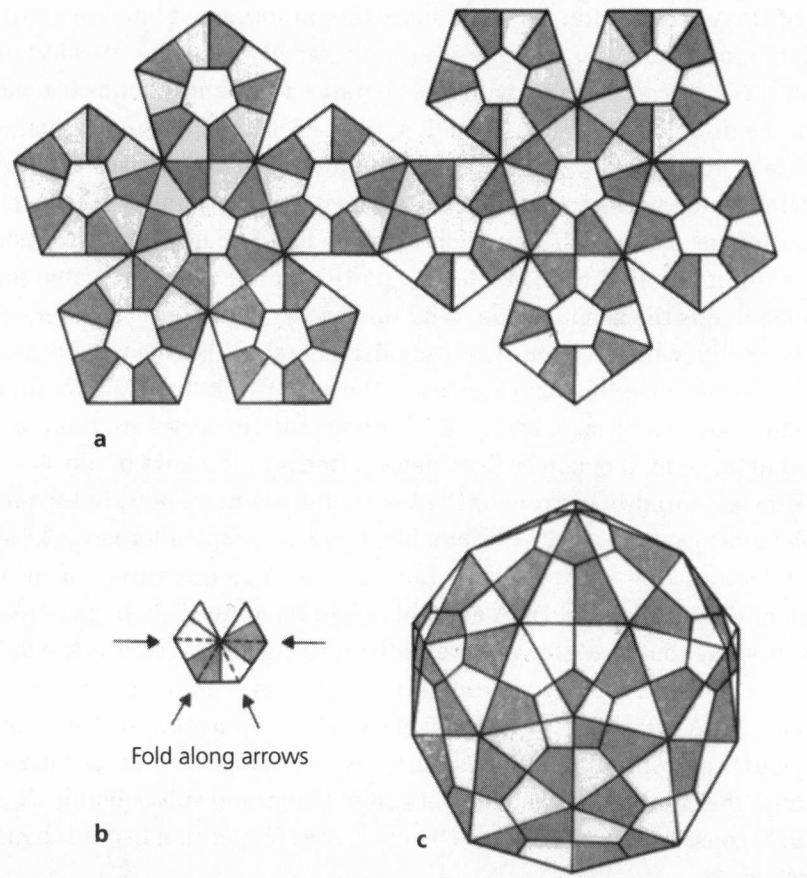

Figure 32
Schematic structure of poliovirus. Glue 12 pentagonal pyramids (b) onto a
dodecahedron (a) to make a three-dimensional model (c).

a special role in the structure of a plant. In fact, magic numbers are the
numbers of identical protein units that can be fitted together in an almost
regular way, to form a nearly spherical surface.[12]

Here's some evidence that viruses know about this constraint. Turnip
yellow mosaic has 32 units; human wartvirus has 72, as do BK virus and
rabbit papilloma. (Human wart and rabbit papilloma viruses are virtually
identical, except that one is the mirror image of the other.) REO virus has
92 units; *Herpes simplex,* the cold-sore virus, has 162 units. The chicken
adenovirus has 252 units, and infectious canine hepatitis has 362
(another magic number) units.

It would be difficult to find more compelling evidence than this patterning of DNA, RNA, and viruses to show the importance of mathematical patterns in making life possible—certainly earthly life, the only kind we know. DNA plays the role that it does because of a simple geometric pattern, the double helix. In a sense, it's more a *logical* pattern, for the key feature is not the helix, but the complementary pairing of bases. It is this pattern, which exists as much in the abstract as in any physical law, that evolution used as a basis on which to create life on Earth. On its foundations, other patterns were erected, in particular the quasi-mathematical enigma that is the genetic code. Why this particular code? Any code will do, basically; whichever one gets established first is likely to win out over any other because life can reproduce. Perhaps Crick was right, and the genetic code is a "frozen accident." Perhaps the Hornoses are right, and the genetic code, too, comes from deep patterns in the laws of physics.

How essential is the role of DNA—not for life here, now, but for life in the most general sense? Presumably, there are many alternative kinds of molecules that can replicate and encode huge quantities of information. Why did we get DNA and not one of the others? Perhaps DNA is the only one that works anywhere in the universe. Perhaps it is the only one that could easily evolve from the chemical mix on the primal Earth. Perhaps DNA is itself a frozen accident—the first replicating-and-encoding molecular system off the blocks, initially imposing itself on our planet because the competition had not yet gotten going, and subsequently dominant because once DNA was in existence, the competition had no chance to get going.

I have no idea. But I do know that without mathematics, we'll never find out.

4

The Oxygen Menace

[I]f the cell acts . . . as a whole, each part interacting of necessity with the rest, the same is certainly true of the entire multicellular organism.

D'Arcy Thompson, *On Growth and Form*, Chapter IV

Earth was changing.

During its first billion years of existence, it had graduated from a spinning mass of liquid rock to a slightly flattened sphere with a solid crust, large patches of open water, and an atmosphere. The first life-forms had appeared roughly 4 billion years ago, piggybacking their way into existence, riding on the back of replicating molecules, or perhaps networks of molecules. They were tiny microbes, precursors of today's bacteria—each one a tiny blob of chemicals surrounded by a membrane. Equipped with malleable heredity, those life-forms had the potential to change themselves—to evolve.

Before they underwent their most dramatic change, however, they changed the Earth. The chemical system that they used to generate energy produced a nasty by-product, a form of toxic waste. Each ancient bacterium, each archaic alga generated only a tiny amount, but over geological periods of time the noxious waste built up. The toxic chemical was *oxygen,* a corrosive, reactive gas that bubbled up out of the primal oceans and accumulated in the atmosphere.

We humans don't normally think of oxygen as being toxic, but that's because our distant ancestors evolved ways to avoid its harmful effects and to exploit it to their own advantage. Even so, oxygen is nasty stuff: It causes rust and fires, and even human divers can suffer from oxygen poisoning. Oxygen has become not just useful but even essential to life on Earth—but it isn't harmless. To those primal life-forms in the early oceans, oxygen was pure poison, and as it built up in the atmosphere, it also dissolved in the sea. Those early life-forms were in danger of drowning in their own wastes.

Some scientists talk of a holocaust in which these early organisms died in the trillions before evolving tricks to mitigate oxygen's toxic effects. This scenario is dramatic, but it is probably also nonsense. It assumes that life behaved like a Looney Tunes cartoon character, walking off the edge of the cliff but not noticing anything until it looked down and discovered that its feet were hovering in midair. Evolution is the result of *constant* adaptation to the environment, not just occasional adaptation. So, even as the oxygen menace was building, Earth's organisms must have been slowly coming to grips with it. Before the menace became disaster, life made a virtue of necessity and turned potential disaster into triumph. An entirely new kind of organism evolved, one that could survive in an oxygen-rich environment: an organism founded on a structure, now known as the cell, which found ways to protect its more sensitive subsystems against the oxygen menace—and to exploit the menace as a source of energy.

The earliest organisms did not segregate their genetic machinery into a special region, the nucleus. Their DNA was just looped around inside a membrane, to which it was loosely anchored. Accordingly, they are called *"prokaryotes"*—Greek for "without a nucleus." Organisms composed of one or more cells that have a nucleus are known as *eukaryotes.* They may be single-celled—the earliest almost certainly were—or the cells may join forces to create a more complex, multicellular creature. The familiar animals and plants of human existence—cats, dogs, cows, frogs, lizards, birds, oak trees, daisies—are all multicellular eukaryotes. If we are to single out any one event that made the greatest impact on this planet and turned it into what we experience today, it has to be the evolution of eukaryotes.

Are there mathematical laws for the behavior of cells? How did cells get together to form what we call "higher" organisms? Are mathematical laws behind those, too? The more closely we look at the cell, the better we understand it, and the more mathematical its operations seem to become. This is true both in general descriptive terms and when we look closely at some of the important molecular structures in the cell, notably its skeletal structure (called a cytoskeleton). Cells move by a dynamic process in which long tubes of a special molecule, appropriately known as "tubulin," are formed, extended, or allowed to collapse and fall apart. Those tubes are extruded by an astonishingly orderly structure, the centrosome, which lies at the heart of every cell. Moreover, the centrosome and its tubes are crucial to cell division. It is ironic: The very kinds of behavior that make cells *seem* so organic and nonmathematical actually have a mathematical origin. Clearly, we have the wrong idea about the capabilities of mathematics. As a result, we may also be misjudging cells.

I said in an earlier chapter that organisms as apparently simple as an amoeba, when seen under a microscope, appear to have a great deal of freedom of movement, of individual volition. I compared them to sand grains borne on the winds. A mathematical approach to the cell is beginning to make it clear that the amoeba, too, is borne on winds; it has no true volition, no freedom of choice. The difference is that an amoeba moves by virtue of a complicated and highly sensitive chemical process, rather than rolling about like a solid lump. The winds that blow it every which way are chemical signals, wafting through its surroundings. It is our failure to observe those chemical winds that drapes the amoeba with the semblance of volition.

Are *we* borne on even more complex winds—winds of emotion, winds of memory, winds of human relationships? Do *we* have true volition? It certainly feels that way to us, but in a sense, we are biased—too closely involved to be dispassionate. I suspect that the answer lies on two levels. On the level of the universe as we perceive it, we have freedom of choice, and we exercise it constantly. On the level of physical laws, however, it may be that our apparent freedom is illusory, a mechanistic response to our own internal state and that of our environment.[1] Because our behavior is an *emergent* property of the physical laws, these two statements need not contradict each other. Right now, such questions are mainly of interest to philosophers—but it may not be so long before they become the domain of physicists and mathematicians. Who knows?

Back to cells. How did they arise? At one time it was thought that the cell arose through a kind of symbiosis among several prokaryote organisms—each apparently "taking in the other's washing," to their mutual benefit. The popularity of this theory slowly declined, mainly for lack of significant new evidence in its favor. Contemporary biological discoveries have not only revived the theory, but also established it so firmly that its overthrow would now require some very surprising discoveries indeed. Meanwhile, mathematicians have been developing increasingly successful models of individual cells and of multicelled organisms and have shown that apparently very sophisticated group behavior of the cells in some eukaryotes can be explained by mathematical equations. One of the most striking examples of eukaryote mathematics can be found in the movement of slime mold, a colony-creature formed from certain kinds of amoebas (which are single-celled eukaryotes). Slime mold forms beautiful spiral patterns, as the individual cells migrate toward a common center; it then turns into a sluglike form, which moves away from its home territory in search of drier ground; and then it puts up a fruiting body on a stalk, which broadcasts spores over a wide area. The spores revert to being amoebas, and the cycle begins anew. It may seem unlikely that such a distinctive sequence of events can be explained by relatively simple mathematics, but since about 1995, mathematicians have found unified explanations for much of the slime mold's behavior.

In fact, the *same* basic mathematical principles of pattern formation govern the geometry of cell division, the form of early stages of embryos, and the strangely beautiful behavior of the slime mold amoeba. Mathematics reveals a common unity among different levels of the living kingdom, a unity that stems from deep universal features of the laws of physics and chemistry.

This unity of cellular life is our story in this chapter, but in order to tell it, I must work my way up from the simple to the complex. I begin with the evolution of the cell. Eukaryote cells are very different from prokaryotes. They are typically 10,000 times as large, measured by volume. Their walls are convoluted, whereas those of prokaryotes are smooth. Most of their genetic material is grouped into highly structured chromosomes kept within the nucleus, though some is found in other regions of the cell, too; in contrast, prokaryotes have just a loop of naked DNA tucked into their wall somewhere, to stop it from rattling around. The interior of a eukaryote cell is divided into many different compartments by membranes, and the cell contains up to several thousand specialized organelles that carry out specific tasks; in a prokaryote, the interior is relatively formless.

The nucleus of a cell is one such organelle; other organelles include the *mitochondria,* which act as energy sources; *peroxisomes,* involved in metabolism; and (in plant cells) *plastids,* the sites of photosynthesis, where energy from sunlight turns carbon dioxide into sugar. The wall of a cell can change its shape. The eukaryote cell possesses a kind of molecular skeleton, made from a protein, the aforementioned tubulin. Unlike our rigid bony skeletons, the cell's cytoskeleton can be rapidly torn down and rebuilt. Indeed, this is how cells move around. Tiny molecular motors run along the tubulin threads, carrying molecules from one place to another and helping in the processes of demolition and reconstruction. If the prokaryote is a chemical workshop, then the eukaryote cell is more like a complex of chemical factories, run by an owner who rejoices in perpetual reconstruction.

Evolution generally proceeds by modifying or combining things that are already in existence. Genuine novelties arise through a gradual process of building, unbuilding, and rebuilding; existing structures are employed as (perhaps temporary) building blocks. What were the building blocks from which eukaryote cells evolved? The answer, most probably: They were the far humbler prokaryotes.[2] The new biological evidence is that, despite their obvious differences, prokaryotes and eukaryotes must be closely related because they share very similar genetics. The evolution of eukaryotes is controversial because no fossils survive to provide clues about what happened. However, the most widely accepted theory—which goes back more than a century but found little favor until it was revived by Lynn Margulis in the late 1960s—is that eukaryotes evolved from prokaryotes by way of a kind of symbiosis. The alternative theories, in contrast, treat the eukaryote cell as a separate, independent evolutionary development.

Symbiosis was first coined with regard to organisms; it occurs when organisms combine forces for their mutual benefit. For example, there is a type of bird that enters the mouth of a hippopotamus and pecks off any leeches that have attached themselves there. This benefits the hippo, who gets rid of leeches that are sucking its blood; it also benefits the bird, who gets food from the hippo's mouth. So the hippo tolerates having a bird inside its mouth and does not close its jaws. The leech, in contrast, is a *parasite:* It benefits from its relationship with the hippo, but the hippo does not.

The theory that Margulis revived holds that the cell evolved as a result of symbiotic interaction among prokaryotes. Those prokaryotes became the eukaryote cell's organelles, and they eventually ceased to have a separate existence, even potentially, as organisms in their own right. The strongest evidence for this theory is that organelles such as mitochondria and plastids possess their own vestigial genetic systems. They have their own DNA, and they have the molecular machinery needed to replicate it and to build proteins from DNA instructions. Their genetic code even differs slightly from that in the nucleus. It is easy to see how this machinery might have survived a symbiotic evolution but hard to see why it would exist if the eukaryote cell were an entirely separate development.

Many of today's eukaryotes can surround and engulf prokaryotes. Indeed, this is how white blood cells deal with invading germs. Therefore, it is reasonable to suppose that the symbiotic partnership arose when some cells acquired the knack of engulfing others—in effect, trying to eat them. Even today, some amoebas pick up passing bacteria and use them as mitochondria, to generate energy—evidence that occasionally an ingested prokaryote might survive for a time, and that its presence might turn out to have beneficial effects. Such an event would be rare, but when it happened, evolution would start to reinforce those effects.

The most likely sequence of events involves a prokaryote losing its (relatively thick) wall to leave only a thin membrane studded with ribosomes—the sites where proteins are assembled. Unrestrained by the thick wall, this membrane grows and becomes very convoluted in shape, with a ruffled edge like the top of curtains—a prototype eukaryote cell. The resulting folds increase surface area without increasing volume, and that improves the efficiency with which the protocell collects nutrients from its surroundings. As a consequence, the protocell grows larger than it would have been able to when it had a thick wall.

At that stage, its digestive processes, like those of prokaryotes, take place *outside* the membrane, in a kind of halo of digestive enzymes. (It's a curious thought—an animal that starts to eat you *before* it actually touches you. . . .) As the protocell grows in size and becomes ever more convoluted, however, it starts to pinch off regions from its outside; these regions form tiny internal pockets. Digestion can then occur inside the protocell, as well as in the external chemical halo. Some such pockets might happen to include the place where the protocell's DNA is anchored; now a rudimentary nucleus has formed, and the protocell has become a genuine cell. At that stage, the cell begins to evolve its skeletal structure, made from long thin molecules that can grow or shrink, according to certain chemical stimuli. This rebuildable cytoskeleton enables the cell to manipulate its outer membrane and to move things about—itself included.

Because the cell can now carry its food around inside it, it no longer has to stay on the surface of a given food source, so it evolves the ability to move around seeking new food, which it can envelop and then digest at leisure. Among the things it ingests are prokaryotes, and some of these are retained because they turn out to be useful. Peroxisomes are probably the first to be kept, because they have the ability to protect the cell against oxygen compounds. Precursors of mitochondria also protect the cell against oxygen compounds and, in addition, can generate an energy-rich molecule, ATP (adenosine triphosphate), which has the ability to store energy and release it when needed. In plants, the final step is to add plastids to the symbiotic society of organelles, so that photosynthesis becomes available as an energy source.

That's the broad evolutionary outline. Now let's take a closer look at what goes on inside a cell because that's where we can see some of the underlying mathematical patterns on which evolution built. For a start, how do cells move? The key to cellular movement is the cytoskeleton, best thought of as an ever-changing web of protein scaffolding, erected inside the cell, propping parts of its membrane up, and leaving other parts to flap around unsupported. Its most important components take the form of *microtubules*—long, thin tubes made from tubulin units. Other protein filaments, notably those made from a substance known as "actin," are also involved, and the different kinds of protein filament interact to provide the cell with its ability to move. Tubulin occurs in two very similar but distinct forms, alpha- and beta-tubulin. Its structure is highly symmetric, like a checkerboard rolled into a tube, with the black squares being alpha-tubulin and the white ones beta-tubulin (Figure 33). Alternatively, imagine building a long, tall chimney from two kinds of brick—black and white—representing the two protein units. At ground level, make a circle of bricks in which black alternates with white. Do the same the next level up, but placing a black brick exactly on top of every white one, and a white brick on top of every black one. Do *not* stagger the bricks as a

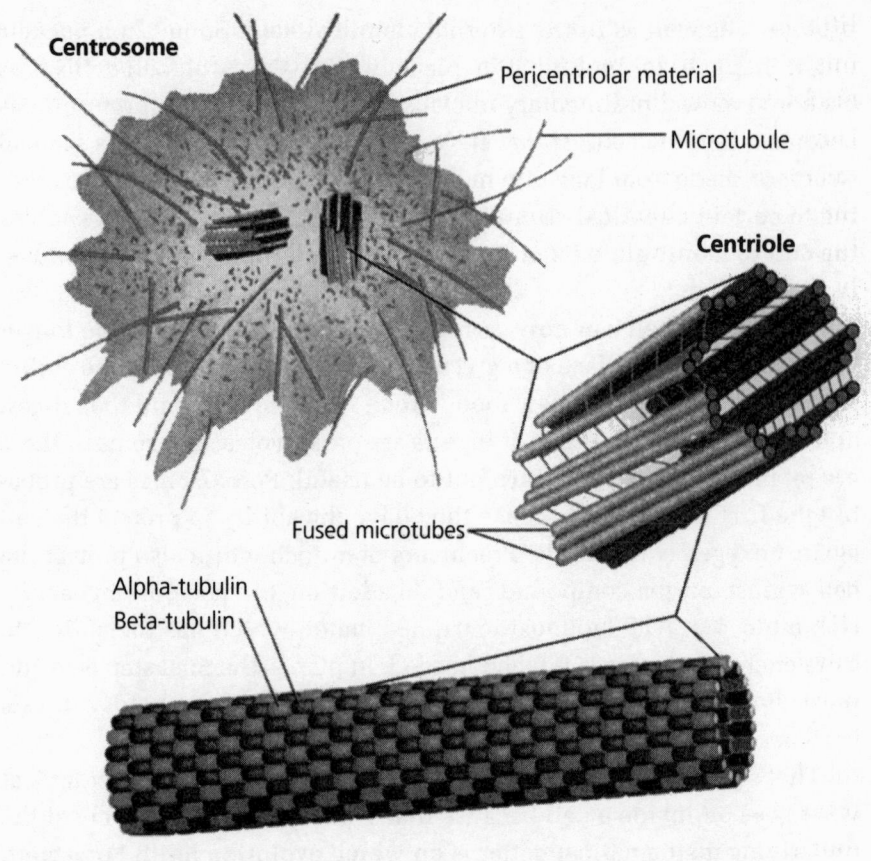

Figure 33
The structure of tubulin, a centriole, and the centrosome.

bricklayer (usually) would, so that each brick bridges the gap between those below. Continue in this manner, alternating black and white bricks, so that the chimney is formed from a series of parallel stacks of bricks.

At first sight, this structure seems to be a silly form for nature to employ. Every bricklayer knows that if you tried to use it for a real chimney, you'd never get very far above ground. It would collapse at the slightest breath of wind. Vertical columns of single bricks, unsupported by others, easily topple. If you made your chimney out of staggered layers of bricks, like normal brick walls, it would be much more stable. So why does nature make such an important item as a microtubule in such an unstable way?

The answer—and it's a mathematical answer—is that in this case, a little bit of instability is a virtue. Because molecules stick together, the unstaggered structure is a lot more stable than one made from nonsticky bricks. Even so, there are long cleavage lines where the structure is weak. However, those cleavage lines turn out to be an advantage. Not only can microtubules grow longer, by adding another layer of protein bricks, but they can also shorten, coming apart at the seams along those cleavage lines—like peeling a banana. In fact they shorten about ten times as quickly as they grow. These two properties let the cell go fishing for interesting things, using tubulin rods, pushing them out at random to see whether they encounter anything, and rapidly collapsing them back and trying again if they find nothing of interest. Thus, it is the unstaggered, marginally stable structure of the rolled-up tubulin checkerboard that equips the cell with the ability to move by building and demolishing its own skeleton. The moral here is that one of the more puzzling aspects of the cell—its ability to move, apparently, at will—boils down to the natural dynamics of a tiny molecular machine with a highly regular structure. Of course, there is much more to cell movement than just a traveling tubulin building site, but that fundamental mechanism is what cellular movement is based on.

What controls the production, construction, and demolition of microtubules? Various chemical signals, some stimulated by the environment: If the cell senses nutrients by detecting chemical signals emitted by the nutrients, then it can respond by tearing down its scaffolding on the side that lies farthest from the nutrients and building it anew on the side that is closest. Mathematical rules of this kind can both propel the cell unerringly toward the most exciting prospects for food and cause it to engulf the food when it reaches the food. Genetics may be needed to set up such a system in a reproducible manner, but it is mathematics—in the guise of physical and chemical laws—that drives the actual movement and makes it possible to begin with. As I said: a partnership.

Of course, something must supply the microtubules to begin with. Genetics? Genes are certainly involved because microtubules are made from proteins, and proteins are encoded in genes. Again, however, there is far more going on than simply genetics, and a lot of it is highly mathematical. Everything depends on one of the most important organelles in any cell: the centrosome, a remarkably mathematical molecular structure, far more regular in form than most parts of a cell. Precisely *why* it is so mathematical is unknown: I suspect that nature made free use of a molecular structure that arose from symmetry breaking. Because that structure happened to produce useful consequences, evolution incorporated it into the structure of every cell.

The centrosome plays a major role in both the formation and control of the cytoskeleton and in cell division.[3] It was first described in 1887 by Theodor Boveri, and independently by Edouard van Beneden, who were both studying cell division in roundworm eggs. When a cell divides, a process known as *mitosis,* the chromosomes replicate and must then be portioned out equally between the two new cells. A structure called the "mitotic spindle" (Figure 34) is central to this process: The chromosomes first line up around the equatorial plane of the mitotic spindle, and they later migrate to its two poles. Boveri and van Beneden observed a tiny dot at each pole of the mitotic spindle, calling them "polar corpuscles" or "centrosomes." When a cell is not dividing, it has only one centrosome, which lives close to the nucleus. When the cell starts to divide, the first thing that seems to happen is that the centrosome splits into two pieces, which move apart. The mitotic spindle forms between them. Therefore,

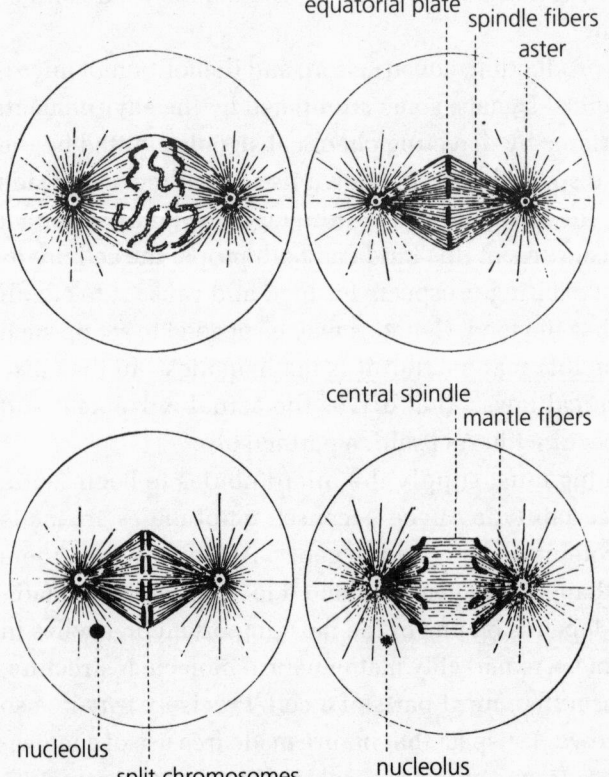

Figure 34
Stages in cell division: The two radiating dots are the centrosomes, and the structure formed by lines between them is the mitotic spindle.

it looks as if it is the doubling up of the centrosome that gets cell divison started and that manages the subsequent stages.

This view was controversial to begin with—as recently as the 1930s, some biologists thought that centrosomes didn't really exist at all but were just artifacts of the way cells are prepared for microscope slides. At that time, it looked as though the cells of many organisms didn't have centrosomes at all, so you can see why they thought the way they did. With the advent of the electron microscope, however, it became clear that all cells have centrosomes, and that their role in cell division really is crucial. Centrosomes extrude microtubules, and during cell division, centrosomes use the microtubules as fishing rods to grab chromosomes and pull them into the required positions.

The centrosome has an elegant, compact mathematical shape. In any animal cell, the centrosome is built around a pair of tiny, symmetric molecular machines known as "centrioles." A *centriole* is a cylindrical bundle of 27 microtubules, arranged in 9 sets of 3, and glued together with a slight twist (Figure 33, right). Two of these centrioles, arranged at right angles to each other, form the core of the centrosome (Figure 33 top, Figure 35). They are surrounded by a fuzzy cloud of "pericentriolar material" (meaning that not much is known about it), from which sprout numerous tubulin fishing rods. By some mathematical mechanism that is not yet fully understood, the two-centriole machinery can encourage the growth of new microtubules.

Rather better understood is the mechanism whereby the centrosome reaches out for chromosomes and pulls them toward it. At the crucial phase of cell division, chemical conditions in the cell make microtubules much less stable than normal, so that they grow rapidly but then shrink again, reaching out at random across the cell. (There are good mathematical models of these processes, derived by the experimental and mathematical physicist Albert Libchaber.) When a tubulin fishing rod hooks a chromosome, however, it stops growing and shrinking at random. Its free end becomes firmly attached to the chromosome, and this stabilizes the molecule—just as an unraveling string can be stopped from fraying by gluing its ends together. The fishing rods have now caught their prey: subsequently, they draw in the chromosomes, using special chemical motors.

The replication of the centrosome is also fairly well understood, at least in outline. Normally, the two centrioles separate from each other, and then each builds a new version of itself, oriented at right angles. At

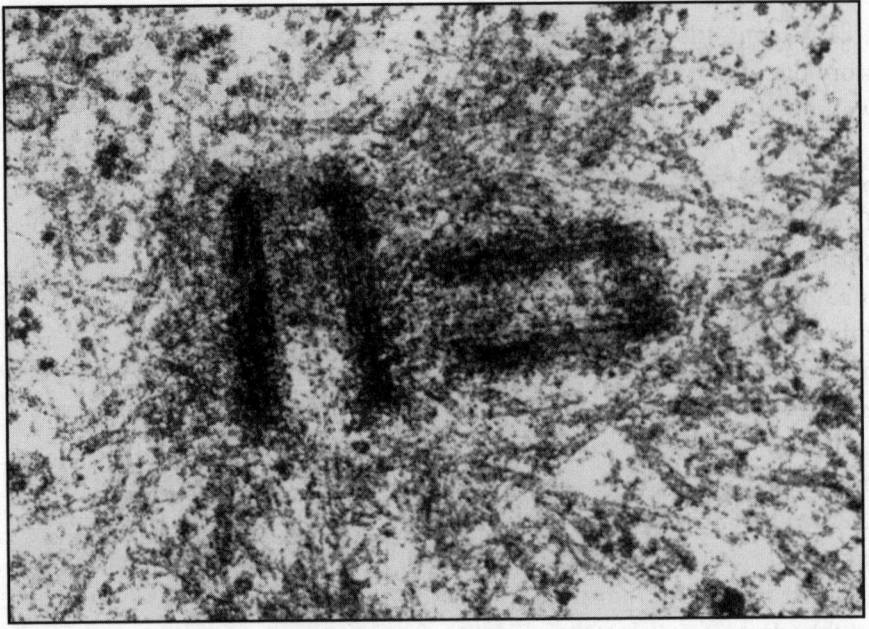

Figure 35
Electron micrograph of the centrosome.

first, the new version consists of only 9 microtubules arranged in a bundle, but soon afterward, it transforms into the normal 27-tube bundle. This process requires an initial centrosome to get the whole thing going; however, it also seems that centrosomes can form spontaneously under certain conditions. In general terms, it is clear that the highly symmetric form of the centrosome must arise by symmetry breaking; it represents a naturally occurring configuration of tubulin units—a kind of minicrystal, constructed from a molecule that forms a tube rather than a lattice. However, nobody has yet carried out detailed calculations to flesh out this broad description.

At first sight, the remarkable dynamics of microtubules seems to deal the death blow to D'Arcy Thompson's speculations about cell division. He argued that the surfaces of dividing cells minimize their energy, just like a soap bubble, and that this general principle governs the overall shape of the cell. It might seem that the existence of the centrosome and its tubulin fishing rods, pulling the chromosomes apart, knocks such speculations on the head. Instead of Thompson's macroscopic physics, cell division proceeds via microscopic chemistry.

However, the separation of the chromosomes is only one part of the story of cell division; somehow, the rest of the cell has to pull apart, as well.

There is evidence that in achieving this division, nature has yet again built on relatively simple physical processes of the same general kind that Thompson had in mind—processes provided free of charge by the mathematical laws that govern the universe. Cell division is a fundamental feature of eukaryote life, and it would be silly to create a vast genetic superstructure to regulate aspects of the process that are perfectly capable of performing adequately by themselves. Instead—as always—the trick would be to tinker with the natural physical processes, subtly altering the context in which they operated, and maybe making the occasional more drastic modification to secure a particularly desirable outcome.

If you watch videos of cells dividing, you will observe a very curious fact: At the time that division occurs, the cell appears *least* lifelike. Lifelike processes involve complex behaviors that look like volition, which result from genetic tinkering to exploit the possibilities that physical laws provide. Instead, during cell division, it is as if the superstructure of complex behaviors must be suppressed, so that this complexity does not interfere with the simple processes of physics that pull the cell, neatly and tidily, into two pieces. In time-lapse photography, you first see astonishingly autonomous cells, irregular in shape, moving with apparent purpose, putting out pseudopods to engulf food—about as nonmathematical as it is possible to be. Then, relatively rapidly, they seem to freeze in place, curl up into round balls, split, and then once more become free agents. Presumably, the process of cell division is too delicate, too carefully organized, to take place when the cell is moving freely. This presumption—if true—adds extra weight to the idea that ordinary physical processes are being harnessed to the task; the more autonomous features of cell behavior could too easily disrupt the process.

There is good evidence that a macrolevel mechanical approach to the form of dividing cells—as opposed to a microscopic view of what is going on inside them in every tiny detail—explains an awful lot of what we actually observe. Surprisingly often—maybe one time in a hundred—cells go astray and divide not into two parts, but into three. It is hard to see how a process orchestrated at every step by genetics could make such a mistake. A mathematical analysis, however, shows that the three-cell state requires only marginally more energy than the two-cell state. Physics seeks the states with the least expenditure of energy, but it is not unusual to get hung up on a state in which the energy is close to—but not equal to—the least possible. That's what seems to be happening in cells.

The cells in a developing embryo divide not once, but many times, and they do so in characteristic patterns. Those patterns are crucial to the early layout of the bits and pieces that will eventually develop into the complete organism. It looks as if the genes trigger cell division, but physics determines where the parts go once division starts. The patterns in which cells divide—their cleavage patterns—display a wide range of geometries, all of which have evident mathematical features, such as approximate symmetry. Here, we focus on just one representative case, the holoblastic radial cleavage pattern (Figure 36). For purposes of description, choose some arbitrary direction, and call it "vertical"; call the plane at right angles to it "horizontal." Then cleavage occurs alternately in the two directions: first a vertical split into two cells, then another vertical one (in a plane at right angles to the first) into four, then a horizontal one into eight, then a vertical split into 16, and so on, alternating horizontal and vertical cleavage planes. It is very hard *not* to detect mathematics in this repeated, regular process of doubling. The natural way to model the overall geometry of the process is to imagine that the cell is a spherical blob of more or less homogeneous sticky stuff— thereby deliberately choosing to ignore the cytoskeleton, all those wonderful organelles, and their like, on the grounds that they may not be terribly relevant to the *geometry* of dividing cells—and to ask what mathematical rules govern division in such blobs. This macrolevel mechanical approach gives the patterns that cleavage is likely to create *if* it occurs. Perhaps the role of the centrosome is to ensure that it does occur, but not to control the geometry because there's no need.

Brian Goodwin tackled this question in the 1980s by studying a "field function."[4] A field function is a quantity that varies from point to point over the cell's surface. For a vivid image, you could represent its

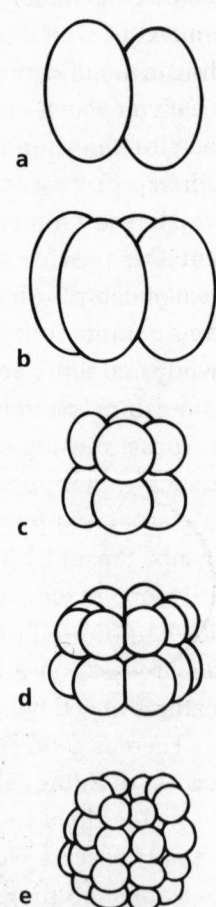

Figure 36
Holoblastic radial cleavage pattern.

values by color coding, so that negative values are represented by coloring the sphere red and positive values by coloring it blue. Different shades of red and blue capture the actual numerical values, but it is the boundary between positive and negative that matters. The coloring determined by the field function varies from point to point, so the sphere is decorated with a pattern in various shades of red and blue. The two colors are separated by boundary curves, and these curves are the most important features. Suppose that the boundary curves determine lines of least resistance to cleavage. If we can model the field function, at least well enough to gain a grip on the geometry of its boundary curves, then we can predict the cleavage pattern. Goodwin did this using a procedure that is entirely analogous to the way that an engineer would model the buckling of a sphere under pressure. The overall spherical geometry leads to celebrated mathematical functions known as "spherical harmonics." The resulting theoretical sequence of cleavages is shown in Figure 37, and it corresponds precisely to what occurs in the actual embryo.

We know that mathematical laws govern the form of the centrosome and the dynamics of the microtubules that it extrudes. Now we see that mathematical laws also govern the overall shapes of cells—but these laws apply on a different level of modeling, one that describes the cell as a whole, rather than listing every distinctive component. The overall form

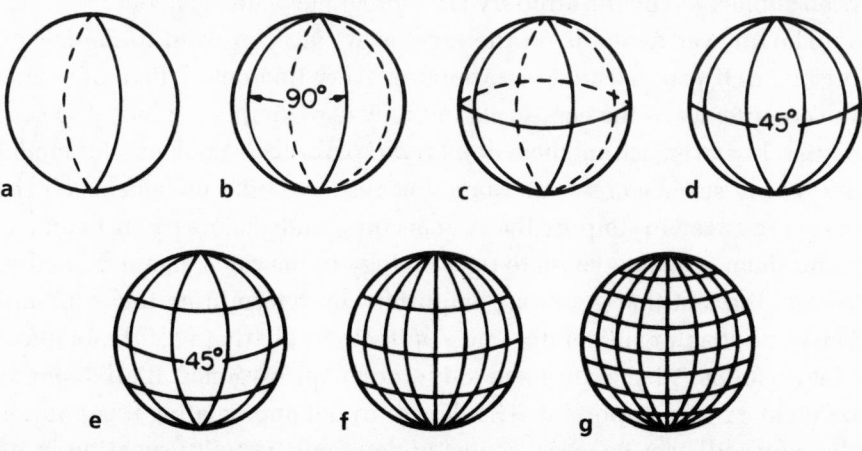

Figure 37

Mathematical model of the holoblastic radial cleavage pattern derived from spherical harmonics.

of the early stages of an embryo is largely governed by mathematics; fine details may well be determined by genetics. We need to distinguish between these two levels of causality if we are to understand the embryo.

The role of genetics, then, is not to determine everything about an organism, but to orchestrate a battery of physical and chemical processes, which, when carried out in just the right way, build that organism. The genes are not a blueprint, in which every component has a well-defined place, clearly marked. They are closer to a recipe. The cell carries out its genetic instructions; the laws of physics and chemistry produce certain consequences, and when you put the two together, you get an organism. When all the ingredients are in the right places, the process works. If anything at all is out of place, problems may arise. Nature discards dozens of failed organisms for every one that succeeds. Its recipes say things like "look on the top shelf of the third cupboard from the left, take out what you find there, and put it into the pan on top of the back right-hand burner," rather than "add two spoonfuls of syrup," and the reliability of those recipes depends—to some extent—on everything else being right.

Regarding cells, and even small collections of cells in a developing embryo, we may be willing to accept that mathematics has a role to play. However, can mathematical models capture anything useful about entire multicellular organisms? How can such models possibly take account of the complicated genetics that is involved in producing an organism and that controls much of its behavior? Hasn't the mathematical structure been submerged by the arbitrary DNA programs of the genetics?

The answer depends on the level at which you want to model the organism. If you insist on incorporatng every tiny detail, then of course the complexity of the genes gets in the way of simple models. As I said earlier, however, astronomers don't try to make their mathematics model every tiny surface crater on Mars. Because even the physical sciences accept the need to simplify the systems they study before trying to understand them, it makes sense to treat biology in the same manner. In other words, the fact that genes are important in determining the form and behavior of an organism does not, *of itself,* imply that useful models of that organism's behavior have to include explicit genes. It all depends on what's really important. The danger in not appreciating this point is that you will ascribe every aspect of an organism to information in its genes. Nobody is silly enough to think that an elephant will only fall under gravity if its genes tell it do so, but the same underlying error can easily be made in less obvious circumstances. Genes supplement the laws of physics—they don't replace them or override them. So whenever you see an organism doing something interesting, you ought to distinguish how much of that behavior, and what part, has a genetic origin, and how much comes solely because it lives in the physical universe and is therefore bound by universal laws.

A good illustration of this point is provided by the humble slime mold—a collective form of the soil amoeba *Dictyostelium discoideum*. Slime mold is an excellent example of an organism for which its behavior could easily be ascribed solely to genetic influences, but its behavior, on close examination, turns out to be based on relatively simple mathematics. The life cycle of slime mold begins (well, cycles don't really begin, so we break into the cycle at a convenient point) as a spore—more precisely, a large number of spores. The spores germinate, becoming individual amoebas, which grow. When the amoebas start to use up the available food, the amoebas begin to move together into a clump. At first, their motion takes the form of circular or spiral waves, but after a while, a treelike structure appears, known as "streaming," as the amoebas flood in toward a rotating central mass. The mass becomes a slug, which moves as if it is a single organism, and migrates to somewhere nice and dry and windy. There, it roots itself, and about half of its constituent amoebas form a long thin stalk, up which the remainder rise to create a fruiting body. The fruiting body disperses a cloud of spores into the surroundings, borne on the wind, and the cycle continues (Figure 38).

The particular behavior that I have in mind happens during the aggregation phase, where the separate amoebas join forces to form a slug. Figure 39 is a typical sequence of observations. In (a)–(c), we see spiral patterns closely resembling those found in the BZ reaction. In (d), the various spiral centers have created separate colonies of amoebas. In (e) and (f), these colonies are continuing to aggregate by streaming in toward the center of each colony, so the gaps among the colonies become more distinct.

The behavior of *Dictyostelium* has attracted the attention of a number of biomathematicians. Here, I concentrate on some mid-1990s work by Thomas Höfer and Maarten Boerlijst,[5] which looks at the early stages of aggregation—spirals and streaming. It turns out that the striking patterns are a consequence of chemical signaling among the amoebas, which produce a chemical known as cyclic-AMP (cAMP: cyclic adenosine monophosphate). The amoebas can sense cyclic-AMP, using certain receptors on their surface, and they respond by moving toward what they perceive as its source. More accurately, the collection of amoebas produces coherent

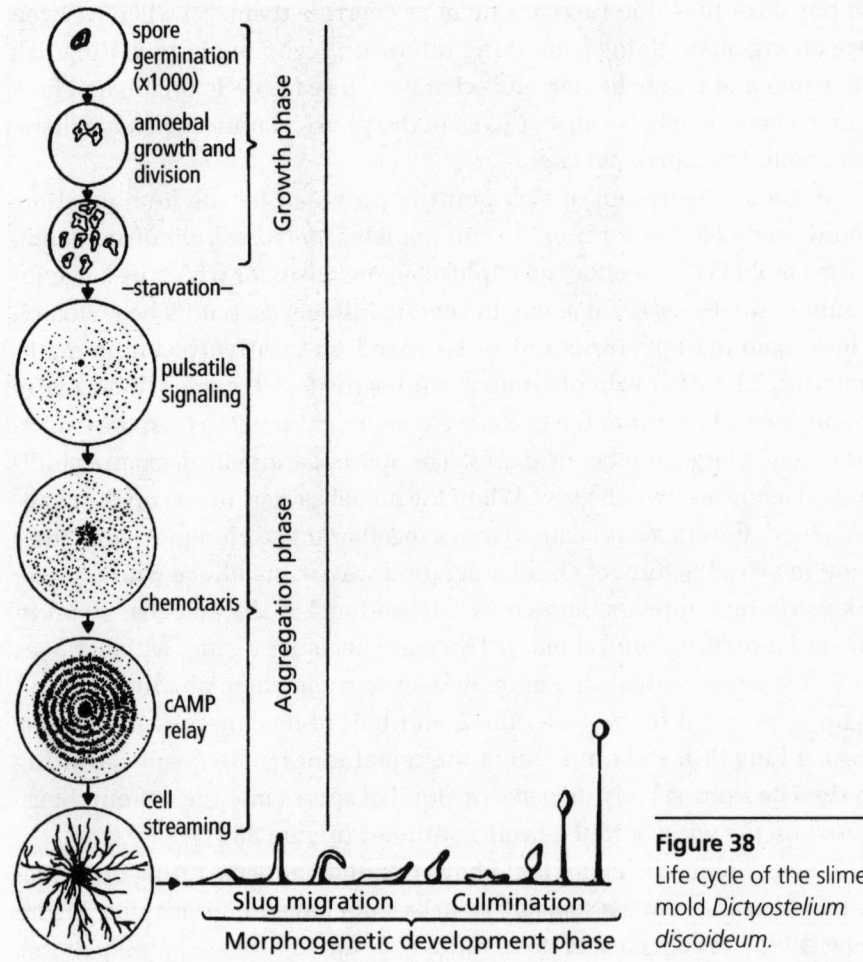

Figure 38
Life cycle of the slime mold *Dictyostelium discoideum.*

waves of cyclic-AMP, and the individual amoebas move upstream as the waves travel past them. It is possible to set up mathematical models of this process, which reproduce the aggregation patterns extremely effectively. One, credited to Höfer, employs just three variables: the density of amoebas, the concentration of cyclic-AMP in an amoeba's surroundings, and the fraction of active cyclic-AMP receptors per cell. Solutions of these equations successfully reproduce both the spiral features and the streaming patterns of the amoebas (Plate 3). Other solutions lead to circular target patterns (concentric rings) instead of spirals (Plate 4); such geometry is also observed in real organisms.

Sometimes the equations go astray and produce streaming patterns with a circular core, like a traffic roundabout[6] (Figure 40a), but that's not

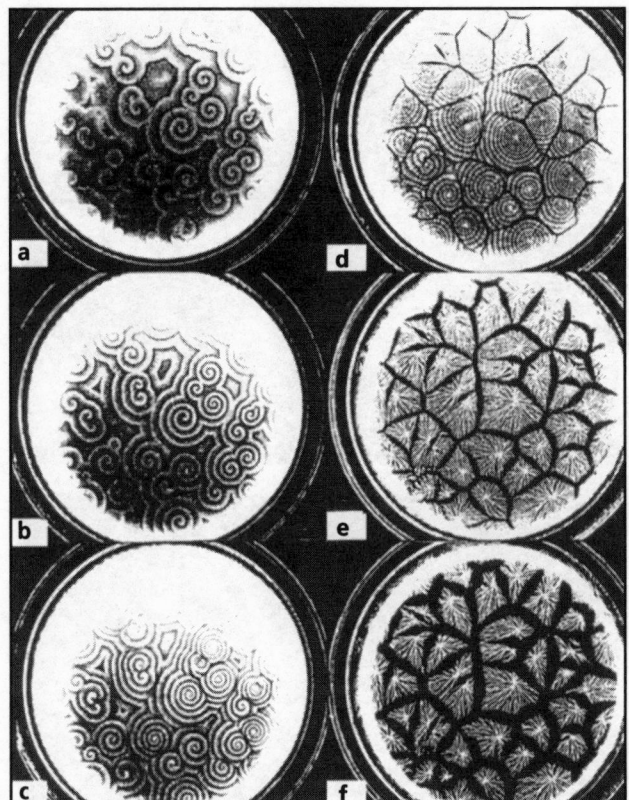

Figure 39
Aggregation patterns in the slime mold.

a problem because the amoebas themselves *also* can go astray, with exactly the same result (Figure 40b). Mathematical equations are especially impressive when they make predictions that at first sight seem to be mistakes, but that turn out to happen in the real world.

The main conclusion here is that a lot of properties of life are turning out to be physics, not biology. Surprisingly sophisticated biological behavior in whole collections of organisms can arise from relatively straightforward mathematical rules. This does *not* mean that biology can be reduced to mathematics. What it means is that the physical patterns that can be exploited by genes often play a far more important role in biology than we tend to think. By using physical processes, rather than just building molecules, genes can construct organisms and control their behaviors, simply and efficiently. I don't believe that we will ever truly understand the role of genetics in the development of organisms unless we bear in mind the effect of physical constraints—an effect that is not

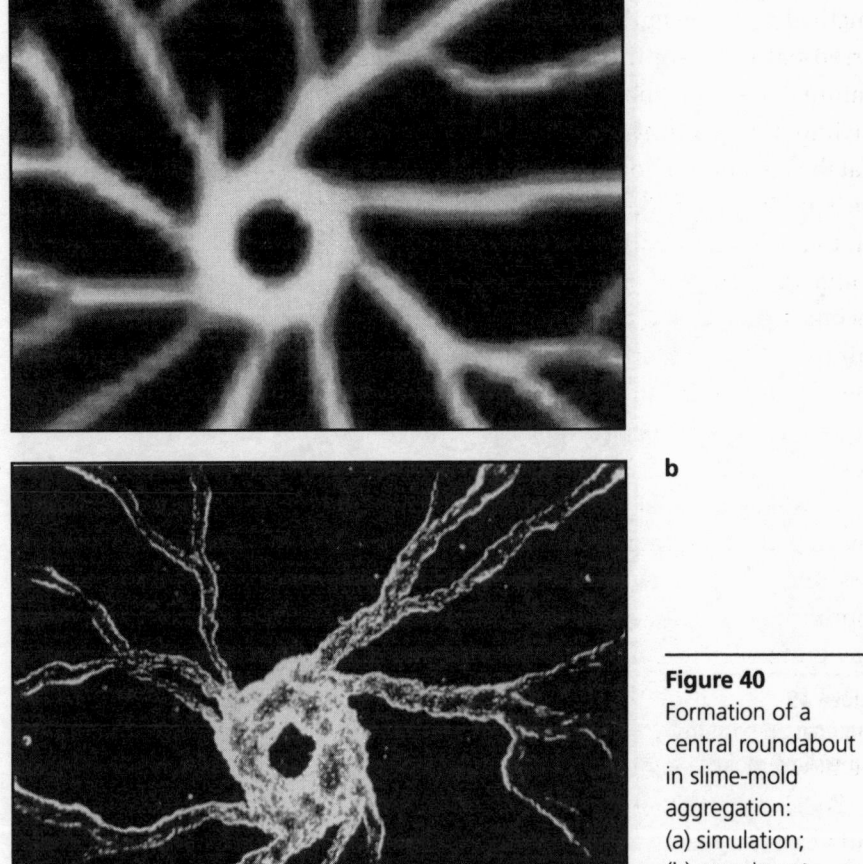

a

b

Figure 40
Formation of a
central roundabout
in slime-mold
aggregation:
(a) simulation;
(b) experiment.

so much limiting as liberating. Leave as much of the details to the physics as possible, then bring on the genes when the going gets tough. *That's* how biological development works.

There is also a deeper mathematical message, to do with the generalities of pattern formation, rather than particular instances. The patterns supplied by mathematics are universal—valid in many different physical systems. For example, the spirals and target patterns in Höfer's model closely resemble those found in the BZ reaction. In part, this happens because Höfer's model involves equations for cyclic-AMP that resemble the usual equations for the BZ chemicals, but there is a deeper reason. The similarity is not just an analogy: Martin Golubitsky, Edgar Knobloch, and I have shown[7] that spirals arise in many different systems for the *same* general mathematical reasons. There is a single, universal mathematical mechanism, and that mechanism has a standard catalog of preferred patterns—in this case, target patterns and spirals. The common feature that selects these patterns is symmetry. For spirals to be generated, the important features are that the system should operate on a flat surface, that the equations should be identical at all points of that surface, and that the equations should not select any direction as being special. Then, early on, there should evolve some distinguished center that breaks the spiral symmetry to that of a circle. After that, the standard catalog of patterns becomes mandatory, and prominent among them are target patterns and spirals. These patterns are specified by classical functions known as "Bessel functions," though with some curious twists that have either been forgotten or were never properly worked out by the classical mathematicians.

To raise the level of generality yet farther, the spherical patterns of cleavage that Goodwin found also belong to a universality class. In this case, the patterns are for systems with spherical symmetry, only now the appropriate functions are spherical harmonics—which, from the abstract viewpoint of mathematics, are close cousins of the Bessel functions that turn up in spiral patterns. In spherical geometry, you expect spherical harmonics to play a substantial role; in circular geometry, you expect to see Bessel functions playing much the same role—and they do.

We started by looking inside the cell, and what we found—flexibility and versatility notwithstanding—was mathematics. Then we looked at how cells divide, and again we found mathematical patterns. Even the peculiarities of entire colonies of organisms turn out to be founded on general mathematical principles—not so much on specific equations as on the common features of entire classes of equations.

What we do *not* see is a world where everything obeys instructions coded in its DNA, and nothing else matters. DNA was the *first* secret of life to be discovered by the human race. Now we must turn at least some of our attention to life's other secret—the universal mathematical principles of growth and form that DNA exploits.

5

Artificial Life

The pointed conical egg of the guillemot is generally supposed to be an adaptation, advantageous to the species in the circumstances under which the egg is laid; the pointed egg is less apt than a spherical one to roll off the narrow ledge of rock on which this bird is said to lay its solitary egg, and the more pointed the egg, so much the fitter and likelier is it to survive.

D'Arcy Thompson, *On Growth and Form,* Chapter XV

Christa Sommerer and Laurent Mignonneau design museum exhibits—but not the dry, dusty displays of rocks and bones behind glass that are the staple of most of the world's museums. Their exhibits are alive—well, nearly. They are computer simulations of strange, lifelike forms, hovering somewhere on the dusky borderland between science and art. They are illusions—but illusions in which the museum visitor can enter and take part.

One of the exhibits is an empty room with a TV camera and a huge projection screen. The camera captures your image and puts you on the screen, which at first is blank. As you move around the room, however, weird plant life appears to sprout around your image. Wave your hand,

and flowers bloom. Walk, and trees sprout from your footsteps. There's only you in an empty room, but on the screen, you are surrounded by dense thickets of vibrant color. For a few brief minutes, you are a god, playing in your own Garden of Eden.

Another of their exhibits is a pool of water, a few centimeters deep. Alongside the pool is a small touch-sensitive computer screen. A child's fingers doodle on the screen—a wobbly squiggle with a blob at one end. In the pool appears a creature, also a wobbly squiggle with a blob at one end.

This one pulsates. It tries to move around the pool. The more hydrodynamic the squiggles, the more effectively its pulsations propel it through the water.

Another child draws a new squiggle—now there are two creatures in the pool. The battle for survival begins. Sommerer and Mignonneau call their computer-generated, optically projected animals "volves," and the exhibit is named "A-volve" (Plate 5).

The A stands for Artificial. Artificial life—not in the Frankenstein sense, you understand: this is not a gallery of monsters. In the sense of "artificial constructs that resemble life." Visitors can design their own volves and set them loose in the exhibit's electronic ecosystem. The computer's programs make the volves react to each other. The system allows for feeding, predators, even reproduction. Volves have their own genetics, a list of information stored in the computer's memory, specifying such features as shape, size, and color. Volves reproduce sexually, mixing their genes at random. They also undergo random genetic mutations.

An overhead camera senses the watching visitors. You can play god and protect your volve with your hand, which keeps the predators away. Like any deity, you then discover that even omnipotence has its limits. A volve that is under protection cannot forage for food, and if you protect your volve for too long, it dies of starvation.

Volves are programmed to look after their children—otherwise their offspring are eaten by predator volves. Most other volve behavior is not programmed explicitly; it emerges from the rules that govern the simulation. The life span of a volve is about one minute. Within the attention span of a human child, many generations of volves are born, compete, and die. The tank is a tiny microcosm of evolution.

A-volve is an artistic construction with a serious scientific background. Similar systems for simulating various features of life, especially its evolution, have created the new discipline of artificial life. Its critics pour scorn on its evident simplifications and the arbitrariness of its rules. Its devotees believe that these are nonissues; they are seeking insight into the universal patterns of evolution, the way it really works. They are not, certainly at this stage, interested in the technical details of earthbound evolution.

They are, however, interested in what makes evolution tick, in what patterns of behavior we should expect from an evolutionary system. What do such systems do easily, what are the genuine surprises? Artificial life is changing our answers to such questions, and by so doing, it is changing the way we look at evolution on our own planet.

Earlier ages saw the rise of the "higher" (i.e., more complex) organisms as an essential feature of the evolution of today's world. The highest organism of all was humanity, and the *purpose* of the whole game was to produce us. Biologists learned, with difficulty, to avoid imputing any kind of purpose or predetermined goal to evolution. On the molecular level, evolution is the result of random changes to DNA. Those changes, realized in the resulting organisms—if there are any, for many mutations fail to lead to viable organisms at all—are then subject to natural selection, and organisms that happen to survive, whether by luck or by good design, get to propagate their genes to succeeding generations. There is no purpose and no sense of direction to evolution—it just does whatever it does.

Artificial life suggests that this resolute exclusion of any kind of overall pattern to evolution is an overreaction. Evolution may not have goals or purposes, for those are human things. Nevertheless, it can have a well-defined direction, a degree of predictability, a dynamic of its own. You can program an artificial life system *knowing* that its mutations are random, that its selection process has no built-in goals, no predefined notion of what is deemed best—and despite this, it will follow a distinctive series of changes, organizing itself into more and more complicated organisms, falling into universal patterns. The very first example of artificial life, Tom Ray's "Tierra," produced—from the simplest beginnings—things such as parasites, social behavior, even a rudimentary form of sex. None of it was programmed in explicitly—but it happened anyway. It evolved.

Many other things that have long puzzled evolutionary theorists are turning out to be completely standard properties of any system that is remotely similar to evolution. One striking feature of the fossil record is *mass extinctions,* in which huge numbers of species die off simultaneously. The best known instance of a mass extinction is the death of the dinosaurs, 65 million years ago, but about 20 possible instances of mass extinction are suspected altogether, and 3 or 4 of them stand out very clearly in the fossil record. At any rate, 65 million years ago, not only the dinosaurs, but also innumerable other species, all died off within what is a very short space of time in geological terms.

Why did this happen? Probably, this particular mass extinction was triggered by the so-called K/T meteorite, which crashed to Earth just off the coast of present-day Yucatán, in Mexico. Other mass extinctions, however, may not have an obvious outside cause. Computer simulations of artificial life have shown that occasional mass extinctions can be the norm, rather than the exception, in many different kinds of evoutionary systems, for reasons that involve only the systems' own internal dynamics. More surprisingly, these simulations have also shown that a tendency for systems to organize themselves into more complex forms may well arise for purely mathematical reasons. If these speculations are even close to the truth, two of the big traditional puzzles of evolution are going to turn out to be based on a complete misunderstanding of how evolution should be expected to behave in the first place.

All of us think we understand evolution. The idea is a simple one. However, the more closely you look at evolution, the more subtle it becomes. For this reason, it will pay us to reexamine some of the usual ground before we return to the exciting, astonishing, but also highly controversial discoveries of the artificial-life brigade.

According to the fossil record, life began with relatively simple organisms and slowly got more complicated. It did so in fits and starts, with occasional bursts of diversity punctuated by long periods of stasis. The reasons for both the bursts and the stasis are hotly debated, with some scientists maintaining that they are what you would expect from a complex system such as life, some appealing to meteor impacts and other catastrophic events, and a few disputing the evidence of the fossil record entirely and denying that either bursts or stasis have occurred. Despite these debates, all biologists are agreed on one overriding thing: the reason why living organisms can change and can pass on those changes to their offspring. The process involved was the brainchild of Charles Darwin, although it was also reached independently by Alfred Wallace. Darwin called the process "natural selection." The *phenomenon* of evolution was already recognized, but not the mechanism behind it, which is what Darwin supplied. Nowadays, we employ the term *evolution* as a catchall describing both the phenomenon and Darwin's theoretical mechanism.

Evolution theory tells us that over long periods of time, species of organisms change. They are not created once and fixed forever; they are mutable. It also tells us why. Darwin came to his conclusions[1] after many decades of studying living creatures. One of his most celebrated examples is "Darwin's finches" on the Galápagos Islands, which lie on the equator, 1,500 kilometers (about 930 miles) to the west of the coast of Ecuador.

The time is 570,000 years ago. The Galápagos Islands are extremely isolated, with no large landmass anywhere nearby. The islands' bird population consists only of seabirds, most of them visitors. The land, with its plants, cacti, hills, and swamps, is inhabited by reptiles—lizards, turtles—but no mammals. It would be a paradise for land birds—except that there aren't any.

Then, by pure chance, a few bedraggled, tired finches arrive, probably blown in by a hurricane. They are all of the same species, and their species had evolved elsewhere to exploit a very precise environmental niche. Perhaps they were ground finches, birds that spent most of the time on the ground, eating grain. Let's suppose, for this example, that they were.

When finches find themselves in land-bird paradise, what do they do? They breed. They enjoy an abundant supply of food, few competitors, and no predators. The finch population must have exploded. Soon there were so many finches that the supply of grain began to run low. There were other potential sources of food: insects, cactus, berries . . . but *these* finches were grain-eaters.

However, the finches weren't identical. They all had ground-finch genes, but some had different genes from others. Some birds with slightly different genes from the main flock, driven to desperation as the grain ran out, found that they could eat small berries instead of grain seeds. Others evolved the ability to eat cactus. As evolution began to work on the now diversifying range of finch abilities, the form of the finches became more specialized. The insect-eaters developed longer, thinner beaks, suitable for catching an insect in flight. The berry-eaters grew thick, short beaks. Within perhaps a hundred thousand years—maybe less—the Galápagos Islands boasted not just ground finches, but also tree finches and birds that bore a closer resemblance to warblers than to finches. And this was just the beginning. So far, that one species of finch has split into 14 distinct species, each with its own lifestyle (Figure 41).[2] Even today, Darwin's finches are still evolving, their genetic makeup and their form drifting gradually as their environment changes.

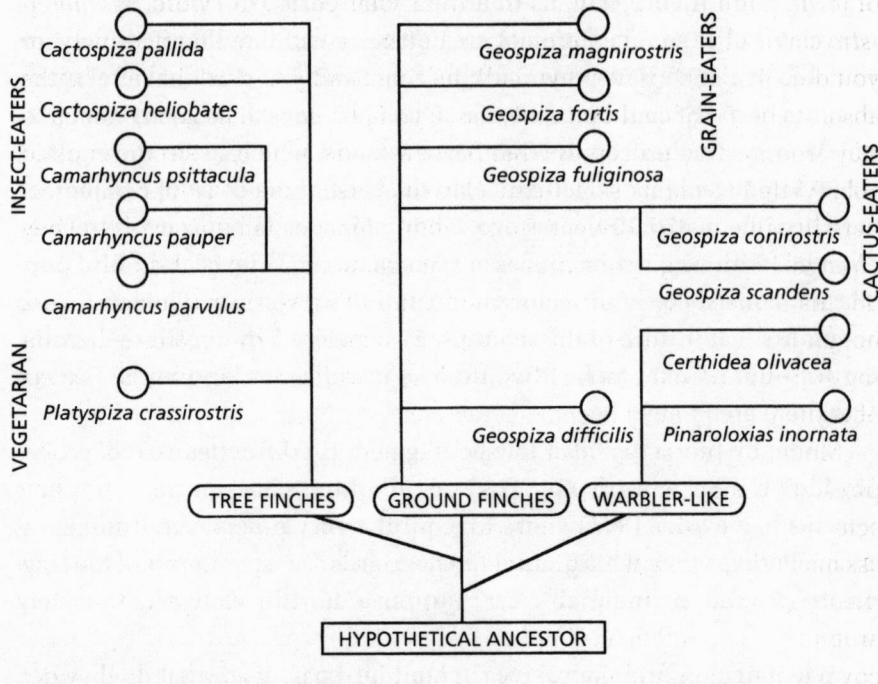

Figure 41
D. Lack's conjectured family tree showing the evolution of Darwin's finches into 14 species.

Darwin realized that something like this scenario must have occurred in the Galápagos Islands because it seemed highly unlikely that 14 separate species of finch could *each* have arrived on a hurricane. The idea that a single species might split into several fitted many other observations that he knew about, so to Darwin, the finches of the Galápagos were the clincher. He worked out the basic mechanism of evolution without having any idea that it was based on random errors in DNA chemistry; that discovery came a lot later. He also realized that evolution involves two very different factors. First, there must be some kind of heredity—parents must be able to pass certain kinds of change on to their descendants. Second, the mechanism of heredity must be slightly imperfect, making the occasional mistake. Given these two properties, everything else follows from the fact that on a finite Earth, all resources are limited. Therefore, organisms will have to compete for resources, and the penalty for losing the competition will be a failure to pass on characters to the next generation. (*Character* is the geneticist's term for any definable feature of form, pattern, or behavior; in ordinary language, the word *character-istic* carries the same connotations.) Either you compete effectively, or you die. You don't have to win all the time, and you don't have to be the absolute best that could possibly exist; you just have to be good enough to stay around. This process is what Darwin called "natural selection," often abbreviated to plain "selection." Darwin's great insight is that imperfect heredity plus natural selection inevitably make organisms evolve. They *change*, becoming better players at the game of survival. Usually, they do this by becoming more complex, but that's a secondary observation, not an explicit feature of the theory, and sometimes they do it by becoming simpler. Also, because all the other players are changing, too, none of them actually need become better at survival *as such.*

Modern biology has filled in one of the big gaps in Darwin's theory, the physical (in fact, chemical) basis of heredity. Organisms pass on characters through their DNA, and errors occur when the DNA copying process makes a mistake. This discovery has come only recently, but a thriving theory of genetics, founded on clear mathematical principles, has been around for much longer. The theory is called "Mendelism," after its discoverer, the monk Gregor Mendel.[3] Mendel was a mathematics student at the University of Vienna. Ironically, he failed his subsidiary botany course and was therefore denied his teaching diploma. In order to pursue his studies, he became a monk, succeeded *too* well, and was promoted to abbot. He was then overwhelmed with administrative duties and had to give up his scientific research. In between, however, he made one of the key discoveries of his century. Mendel kept lots of pea plants, and he cross-bred them, pollinating plants with the pollen from other plants. He found that plant heredity displayed simple numerical patterns—for example, when he cross-bred green-seeded pea plants with yellow-seeded ones over multiple generations, he got three times as many yellow-seeded offspring as green-seeded ones. From such results, he deduced that the factors determining the characters of the plant must be inherited from *both* parents. Nowadays these factors are termed *alleles;* these are related to, but distinct from, *genes,* a word that has made its way into everyday language. Alleles are the different forms that a gene can take. For instance, the gene for seed color has at least two alleles, yellow and green.

The tidy numerical ratios were vital clues to life's genetic mechanisms. As an example, let me explain that 3:1 ratio in peas. Mendel's idea is that the parents each have have *two* alleles, and the offspring inherits one from each of them—chosen at random. Call the alleles for seed color Y and G, so that the possible pairs are YY, YG, GY, and GG. If a pea plant has alleles YY or GG, then it is clear what color its seeds should be; but what about YG or GY? Mendel's answer is that in such cases, one particular allele always wins. That allele is said to be *dominant,* the other *recessive.* In peas, Y is dominant and G recessive, so each of the pairs YY, YG, and GY leads to offspring with yellow seeds, and only GG leads to green. Note the numbers: three pairs yellow, one pair green—the magical 3:1 ratio.

A huge amount of mathematical machinery for handling this kind of calculation was developed in the early twentieth century by the statistician Ronald Aylmer Fisher. Its virtues are simplicity and pencil-and-paper accessibility. Its main defect is that it employs a simplifying assumption: Large numbers of different individuals are homogenized into a common gene pool and it tracks only the frequencies with which alleles occur, not who has which alleles and in what combination. Geneticists see evolution in completely different terms, compared with Darwin: Instead of highlighting organisms and characters, they focus on genes and alleles. Organisms are a kind of secondary by-product of genes; only genes really matter. This point of view, which dates from the 1930s, is called "neo-Darwinism." In the modern era, that focus has been sharpened still further, and the source of all important action is seen as the molecule DNA.[4]

Moreover, the study of evolution in terms of DNA has become highly mathematical. There are regularities and patterns even in the random mutations of DNA bases, and we can use them to trace evolutionary histories. Admittedly, the patterns are mostly statistical, and the whole area is rather controversial, for good reasons. As always, the mathematical models are only as good as their assumptions, and it now looks as if the earliest work was a bit naïve in that regard. I don't consider this a major criticism: Every new idea has to start somewhere, and pioneering work is always naïve in retrospect. Naïve or not, it's a fascinating story, and it made radical changes to our view of evolution—including our own evolution.

The central idea is to employ precise mathematical techniques[5] to trace evolutionary histories—the jargon is "phylogenies." Before the subject went mathematical, phylogenies were constructed on the basis of expert opinion—the gut feeling that a particular species of beetle, let us say, was evolutionarily close to another beetle species, but more distant from centipedes or wasps. The problem with such methods is that experts can disagree, and there is then no way to resolve the dispute rationally. Mathematical methods held out the hope of being more objective because one of the great advantages of mathematics is its precision. Unfortunately, this can also be one of its great disadvantages because precision is not the same as accuracy. For example, a tape measure can tell us a person's waist measurement with a precision of 1 millimeter (a twenty-fifth of an inch) or better. However, the accuracy of this measurement depends on how tightly we pull the tape. Similarly, the accuracy of a mathematical answer is no better than the assumptions on which it rests—but it is easy to be so impressed by the precision that you don't question the assumptions.

There are two basic types of methods for tracing the family trees of organisms. One type is to deduce relationships by looking for common characters—for instance, all birds must be related because they have wings and feathers; bats have wings, too, but no feathers, so they are more distantly related to birds. The other type of method is to ask, given two organisms, how far back in evolutionary history did they diverge from a common ancestor? The first approach, generally called "numerical taxonomy" and pioneered by P. H. A. Sneath and R. R. Sokal, involves making a list of characters—shape of bones, pattern of veins, banding pattern of chromosomes, whatever. Then these characters are assigned numerical values. For instance, suppose we wish to distinguish among a hippopotamus, a fly, and an ant. We might draw up a table of such values, like this:

CHARACTER	HIPPO	FLY	ANT
body length (cm)	375	2	1
number of wings	0	2	0 (mostly)
number of legs	4	6	6
lives in water?	1 (yes)	0 (no)	0 (no)
and so on.	. . .	. . .	. . .

The problem now is to extract from this list some quantitative measure of the overall differences among the various creatures. One way is to represent each animal's list of character values as a point in a multidimensional space, and to see how the points cluster. To keep the idea simple, focus only on the first two characters in the list: length and number of wings. We can represent those characters graphically (in "morphospace"), using two perpendicular axes, and we can locate the three animals in the resulting two-dimensional space (Figure 42). Visually, it is clear that the

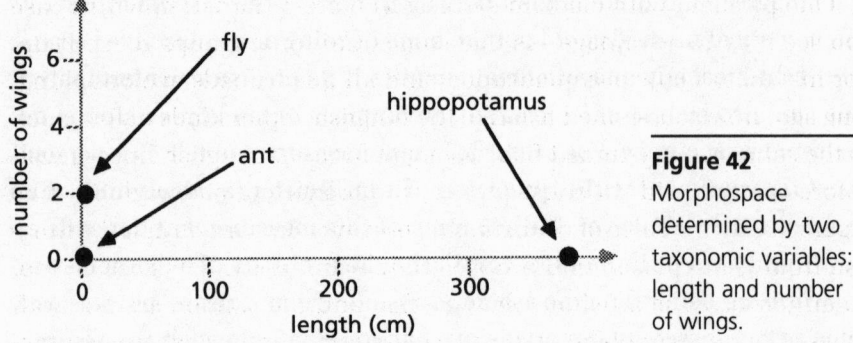

Figure 42
Morphospace
determined by two
taxonomic variables:
length and number
of wings.

ant and the fly are closer together in the picture than either is to the hippopotamus. A great deal of mathematical technique has been developed to make sense of this kind of clustering information in spaces with large numbers of variables, and to make the process as objective as possible.

However, it is not as objective as its proponents claim. The *calculations* are indeed objective, but the *assumptions* underlying them involve human judgment as to which characters really matter—and indeed just how to define the distance between representative points. Again, people confused precision (undeniable) with accuracy (contentious).

Critics of this taxonomic type of approach also point out that it has no evolutionary content. In contrast, their alternative phylogenetic methods try to work out how recently two given species diverged from a presumed common ancestor. The phylogenetic type of approach can change the resulting picture dramatically. For example, when applied to these three organisms—lobster, barnacle, and limpet—numerical taxonomy concludes that the barnacle and the limpet are the most closely related of the three, whereas phylogenetic methods place the lobster closest to the barnacle. There are two main phylogenetic schools: *evolutionary taxonomy,* promoted by Ernst Mayr, George Gaylord Simpson, and Theodosius Dobzhansky; and *cladism,* introduced by W. Hennig. Both schools consider the lobster a closer relative of the barnacle than the limpet is. In its original form, cladism was also based on measurements of characters, but it sought to deduce the *real* evolutionary tree, or lineage of the organism—its list of ancestral organisms or species. It therefore looked at characters shared by groups of organisms that were presumed to be evolutionarily related, and it focused on only those characters that were unique to some group. Elaborate mathematical techniques were devised to locate such characters and to deduce the family tree.

One problem with cladism—unless you are a cladist, in which case you see it as an advantage—is that some traditional groups get split up. For instance, a cow is a quadruped, and all quadrupeds evolved (long, long ago) from lobe-finned fish like the lungfish. Other kinds of fish, such as the salmon, a ray-finned fish, diverged from the lungfish lineage well before the cow did. In consequence, the branching pattern looks like Figure 43. By the rules of cladism, this prevents the salmon and the lungfish from being placed in the same group—unless the cow goes in, too. So either a cow is a fish, or the lungfish and the salmon are not both fishes. The same problem crops up in classifying reptiles, where the crocodile is found to be a closer relative of the bird than of the lizard or the snake. The cladists' viewpoint is that the family tree is just like that: too bad. The numerical taxonomists disagree, which is where the third school, that of evolutionary taxonomists, comes in—with a compromise approach that pleases neither of the other two schools.

Nowadays, however, we have an (allegedly) less contentious way to trace evolutionary lineages. Instead of tracking characters, we track DNA codes. If one organism has a sequence somewhere that goes CCGGGTTTCC, and another has a sequence in the corresponding place, which goes CAGGGTTTCC, with only one variation, then the two must be more closely related to each other than to one with the sequence CGTGACTTCC, which differs from them both in many more positions. There are still some big surprises, but the evidence is distinctly less subject to personal bias in choosing characters. DNA isn't the only molecular method for tracking lineages; for example, amino acid sequences in proteins can also be used. The biggest problem here is a rather interesting mathematical one: How do you define a sensible distance between DNA sequences? The obvious one is what communications engineers call the "Hamming distance": the number of places where the sequences

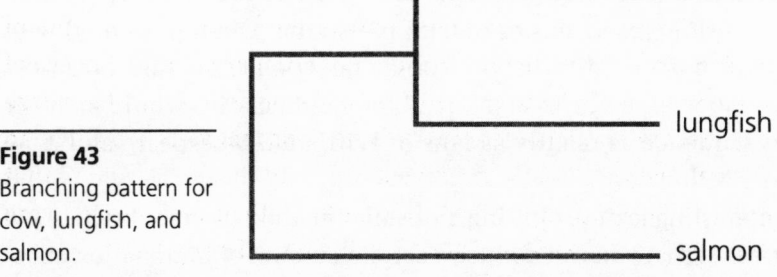

Figure 43
Branching pattern for cow, lungfish, and salmon.

differ. So the Hamming distance between

CCGGGTTTCC

and

CA**G**GGTTTCC

is 1, because there is only one different base (boldface), whereas that between

CCGGGTTTCC

and

C**GTGAC**TTCC

is 4. Unfortunately, DNA does not always mutate by just changing one base. Bases may be inserted or deleted; so may whole sequences of bases. Sequences can be copied several times in a row. Moreover, sequences can also be inverted. So only one step away from CGTGACTTCC, we find sequences such as

(insert a base) CG**A**TGACTTCC
(delete a base) CG GACTTCC
(insert a sequence) CGT**ATTAG**GACTTCC
(delete a sequence) C GTTTCC
(double up a sequence) CGTGAC**TGACT**TCC
(invert a sequence) CG**TTCAG**TCC

The Hamming distance places all of these a long way away from the original sequence. You might decide to use something like "the smallest number of such operations that can transform sequence 1 into sequence 2." Apart from being very hard to calculate, this measure also has a major defect. In a literary analogy, it would place *Winnie-the-Pooh* very close to *Hamlet*. Two steps alone separate them:

Step 1. Insert the whole of *Hamlet* at the end of *Pooh*.

Step 2. Delete *Pooh*.

So that won't work either. What is really needed is a way to characterize *sensible* insertions or deletions—so that when presented with "TO BE OR NOT TO BE, THAT IS THE POOH TRAP FOR HEFFALUMPS QUESTION. WHETHER 'TIS NOBLER IN THE MIND . . . ," the mathematics would spot the inserted sequence as readily as fans of William Shakespeare and Alan Alexander Milne do.

An interesting feature of using molecular methods to trace branches of family trees is that you can deduce a certain amount of information about when particular species diverged from others. The idea is that particular regions of the genome mutate at different rates, and that mutation rates can be estimated from modern experimental data. So, in effect, the DNA mutations provide a molecular clock. There is a certain amount of disagreement about just how *regularly* the clock ticks, but on a qualitative level, the idea is sensible enough. A great triumph of this approach occurred in *physical anthropology,* the study of humanity's prehistoric ancestors. Until the 1960s, the fossil species *Ramapithecus* was widely considered to be a hominid—a very close relative of humankind, much closer than the great apes such as the gorilla and the chimpanzee. In 1967, however, V. Sarich and A. C. Wilson[6] measured what they called the "immunological distances" among humans, gorillas, and chimpanzees, by seeing how strongly antiserums from one of these species bound to the protein albumin in the others. Their results, interpreted via the molecular clock, indicated that humans diverged from the great apes only 5 million years ago. Other evidence showed that *Ramapithecus* and humanity diverged more than 9 million years ago. So *Ramapithecus* wasn't a hominid after all.

That figure of 5 million years has now come into dispute—though not the conclusion that *Ramapithecus* wasn't a hominid. In March 1997, Simon Easteal and Genevieve Herbert took another look at the ticking of the molecular clock. The figure of 5 million years arises from an estimate of 1.5×10^{-9} for the probability that a given base will mutate in a given year. (This means that any given DNA base will, on average, mutate once every 600 million years. Mutations in any given base are *very* infrequent—but there are an awful lot of bases.) Easteal and Herbert argued that the mutation rate ought to be pretty much the same in all mammals, but that assumption placed the divergence of the marsupials (such as kangaroos) from the other mammals at about 330 million years ago. However, fossil evidence shows conclusively that the divergence occurred no more than 125 million years ago. The two scientists concluded that the molecular clock ticks about 50 percent faster than had previously been assumed. This led them to revise the date at which humans and chimps diverged— it was probably 3.6 to 4 million years ago, not 5 million. This revision of history is important because it makes it possible for a known hominid, *Australopithecus (A.) afarensis,* to be the common ancestor of both chimps and humans. Another similar hominid, *A. africanus,* could then be the ancestor of gorillas. I mention how scientific understanding has developed mainly to show that science continues to refine its understanding of early human evolution, and that mathematics is proving an indispensable tool in physical anthropology.

Yet another method for estimating molecular distances is one known as DNA hybridization. This technique involves mixing DNA strands from two species, in solution, where they bind together, and then using the synthesis points of the combined strands to estimate how strongly they are bound. The more strongly they are bound, the more closely the DNA sequences match. By this method, C. G. Sibley and J. E. Ahlquist[7] showed that we are more closely related to the chimpanzees—both ordinary chimps and bonobos—than to gorillas (Figure 44). The *bonobo* is a species of chimpanzee that was recognized only recently; it is less robustly built than the chimps you see in TV advertisements and most other media.

Evolution raises some novel problems for mathematics because when viewed as a process, it has unusual features that do not fit neatly into existing mathematical theories. Evolution has at least four ingredients:

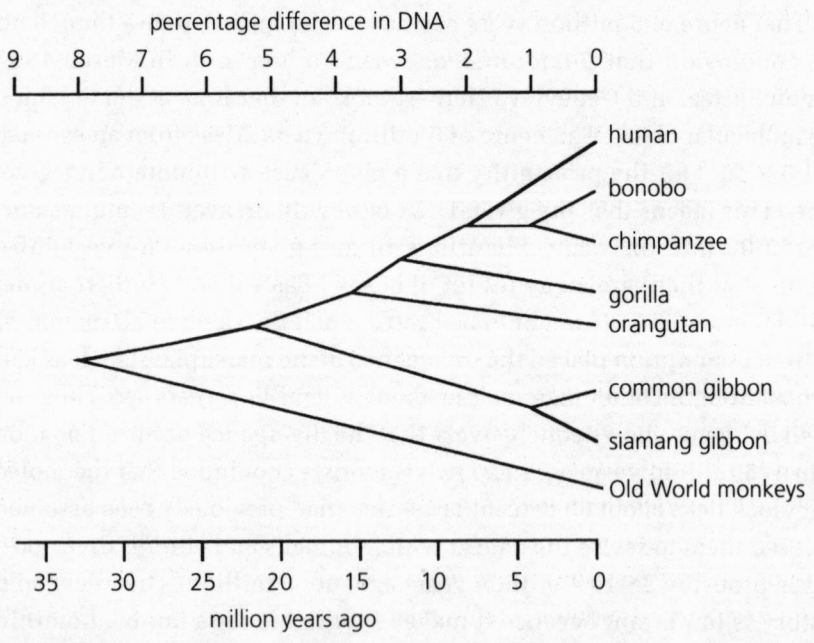

Figure 44
Evolutionary family tree of human beings, as deduced from DNA hybridization experiments.

◇ Mutation

◇ Selection

◇ Development

◇ Environment

These ingredients interact to produce organisms that are adapted to their environment. Genes affect organisms by controlling—or at least influencing—the organisms' development. Genes themselves change by random mutation. These processes involve the content of each organism—its own internal structure. Organisms affect the range of genes available in the next generation—the gene pool—by reproduction. Selection affects organisms by favoring those that are well adapted to the given environment. These processes involve the organism's context—the environment, including other organisms, climate, terrain, and availability of mates (in sexual species). Evolution occurs when many organisms engage in these interactions, and more or less systematic changes emerge. A fully realistic model of evolution must take all of these interactions into account—a daunting task.

Geneticists, especially those known as neo-Darwinists, try to sidestep the messiness of organisms by collapsing the evolutionary system down to something simpler, looking only at the effects experienced by genes.[8] The complex processes of selection within a changing environment are collapsed down to a single fitness factor for a given allele; and phenotype—the form and behavior of the organism—is assumed to be a direct consequence of genotype. Instead of organisms competing for the right to reproduce, neo-Darwinists see genes competing for their place in the gene pool. Moreover, in classical genetics of the kind introduced by Fisher, a heterogeneous ecosystem—such as a rain forest full of diverse plants, insects, small animals, and predators—is modeled as a homogeneously stirred pool of genes. As the organisms reproduce, those genes get mixed together in new combinations; as natural selection weeds out unfit alleles, the alleles that allow organisms to survive better tend to proliferate. Random genetic mutations keep the gene pool simmering. The mathematics focuses solely on the proportions of particular alleles in the population, and it models how those proportions change in response to selection. Physicists call this kind of approach a "mean-field theory," and they resort to it only when desperate. In *mean-field models,* a collection of distinct individuals is replaced by a homogeneous mass of identical average individuals. It's like assuming that every family really does have

2.3 children—fine for some purposes, such as deciding how many schools to build, but misleading for others, such as deciding how many big or small houses are likely to be needed in the next decade.

For example, a hypothetical population of slugs might have genes for green or red skins, and other genes for a tendency to live in bushes or in bright red flowers. Typical genomes include green/bush and red/flower—four possible combinations altogether. Some combinations, however, have greater survival value: for example red/bush slugs would be easily seen by birds against the green background of the bushes they inhabit, whereas red/flower slugs would be less visible. To model this system in the spirit of Fisher, we assign numerical weights, called "selection coefficients," to the possible genomes. Thus, red/bush might have a selection coefficient of 0.1, compared with 0.7 for red/flower. Essentially, these choices indicate that a red slug living on bushes has only a 10 percent chance of surviving to reproduce, whereas a red one living in flowers has a 70 percent chance. We also assume some initial distribution for the proportions of the total slug population that correspond to each of the four pairs of alleles—say that 20 percent are red/bush, 15 percent red/flower, and so on. Fisher's mathematical scheme then lets us calculate the proportion of each allele in each succeeding generation. If some proportion becomes zero, then that particular allele dies out.

All this, of course, is no better than the assumptions that go into it—which, by today's standards, are unsophisticated. As well as being mean field, Fisher's genetic models are also *linear*—they assume that the effect of an allele is proportional to the frequency with which it occurs, and that the effects of different alleles simply add up. Linear mathematics held sway in classical times because the calculations were simple enough to be done with pencil and paper. Today, most areas of science are adopting nonlinear models, with more complex, but far more realistic, dynamics. The same is true of frontier genetics and evolutionary theory.

We can capture some of the flavor of nonlinear modeling by using a geometric analogy. Imagine a plant on the side of a hill, producing seeds and scattering them randomly around it. Suppose, for the sake of argument, that seeds landing higher up the hill are retained, but those landing lower down are removed. Then, over a period of time, you will find a patch of plants working its way higher and higher up the hill. This image of a hill is a simple example of what Sewall Wright[9] called a "fitness landscape." Such a landscape is a surface graph (Figure 45) that represents how the fitness of an organism depends on its characters. Fitness is

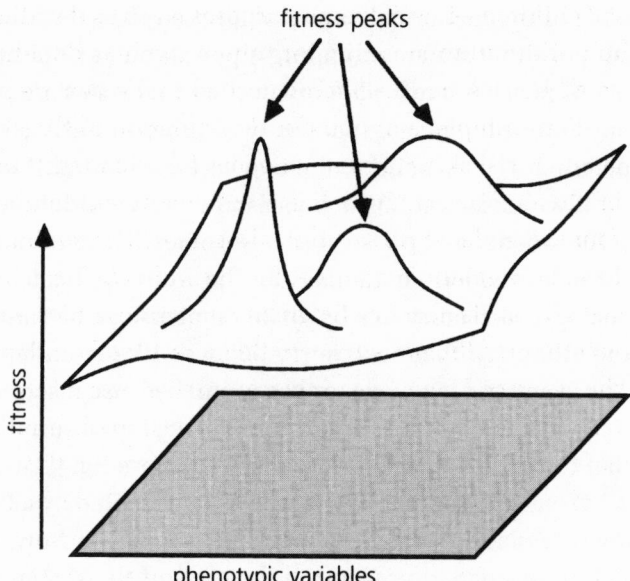

fitness peaks

fitness

phenotypic variables

Figure 45
Sewall Wright's
fitness landscape.

represented by height, and the characters determine the organism's position on the landscape. Nonlinearity implies that a typical landscape is bumpy, rather than being flat or just sloping at a constant angle. The bumps are the places where organisms are fittest, so they dominate evolutionary behavior; the valleys are also important, however, because they separate the bumps from each other.

Fitness is a relative concept, not an absolute one, but it's clear what such a model predicts: Organisms will evolve in the uphill direction, heading toward the local peaks of fitness. There are dozens of ways in which this model of evolution is too simple to capture the reality in detail, but it emphasizes the basic point: Even if the errors in heredity are random, natural selection will impart a definite directionality toward fitter organisms. Similar models can cope with more realistic assumptions, and they lead to much the same conclusion.

Wright's image was one of the earliest biological instances of a way of thinking that has now become all-pervasive, the mathematical concept of a *phase space*. This idea, introduced by Henri Poincaré a century ago, represents dynamics as geometry. A *phase space* is a multidimensional mathematical space, the points of which represent possible states of some dynamical system—a system in which its state can change over time. In Wright's model, the phase space is organism space—its coordinates form a list of the numerical values of all the relevant characters. For example, suppose that we are modeling a population of finches. Then the system is "all possible finch phenotypes," and its states are *particular* finch phenotypes. If the phase space is two-dimensional, a plane, its two coordinates may correspond to the two variables "height" and "wingspan."

If we looked at 20 such variables, we would, in effect, be working in a 20-dimensional phase space—a somewhat mind-boggling concept that pervades modern mathematics. The word *dimension* is introduced as an analogy; each entry in a list of 20 numbers can be varied independently of the others, so that each entry behaves like an independent dimension. The geometric language proves useful because it sets up helpful analogies with spaces of 2 or 3 dimensions; precision is supplied by remembering that the actual objects under discussion are long lists of numbers.

What of dynamics? Dynamics is represented by a flow pattern in phase space. As a representative point goes with the flow, its *coordinates* (the list of numbers representing the state of the system) change over time. Phase spaces offer technical advantages for mathematicians, but their most important role is metaphorical: They formalize the notion of context by embedding what *actually happens* in a structured realm of all the things that *might have happened* instead.[10] In a phase-space model, you can ask "why *this* behavior rather than *that?*" and hope to get a sensible answer. Wright's phase-space approach immediately solves one worrisome puzzle: If phenotypic variables are *continuous*—capable of assuming any numerical value within a given range and thus changing gradually—why do we see well-defined species with values that cluster around particular numbers, and empty gaps at other numbers? The reason is that continuously varying landscapes still have *isolated* peaks.

The main defects of Wright's approach are these:

◇ Not all characters can be captured by continuous numerical variables.

◇ Fitness is not just a matter of evaluating a single number. (For example, in relation to a goldfish, a cat is fitter at climbing trees, but less fit at staying underwater for hours at a time. Which animal wins depends on what game they are playing.)

Nevertheless, the fitness landscape is a useful and insightful metaphor for certain aspects of evolution. In the late 1960s, Christopher Zeeman developed some ideas of René Thom and set up a mathematical model of the fitness landscape, which addresses the slippery issue of whether evolution can jump.[11] Can a radically new organ (such as an eye) or even a radically new species *suddenly* come into being? Darwin's view on this question is generally presented as unequivocal: *"Natura non facit saltum,"* "nature does not make jumps." Also, however, he said, "Many species once formed never undergo any further change . . . and the periods during which species have undergone modification, though long as measured by years, have probably been short in comparison with the periods during which they retained the same form." That is, the jumpiness of evolution depends on the time scale over which you look.

It is true that every so often, the fossil record seems to show some sudden changes. Are the apparent jumps genuine, or do they just reflect gaps in the fossil record? Certainly the fossil record is incomplete, and apparent gaps are still being filled at a relatively rapid rate, as paleontologists dig up new specimens. The gradualist view of speciation is that over long periods of time, the phenotype of a species slowly drifts, until eventually the change becomes so great that the species seems to have changed. For example, the trilobite, which used to live on the ocean floor, gradually evolved for hundreds of millions of years. As new species arose, earlier ones died out. All of the known changes are arguably gradual in character; then, about 250 million years ago all trilobites died out altogether.

In 1972, Niles Eldredge and Stephen Jay Gould[12] caused a storm of controversy with their theory of "punctuated equilibrium." This theory maintained that (1) speciation nearly always occurs by the splitting of lineages, not by the slow drift of a single lineage; and (2) this splitting happens much more rapidly than does the usual drift rate. They tied this theory (unnecessarily, it seems to me) to the prevailing idea that the splitting of species occurs by *allopatric speciation,* in which a small subpopulation on the edge of the geographical range moves away and becomes disconnected from the main body of the species. Once isolated, this group evolves in new ways because it is in a different environment. If the resulting new species reinvades the original territory, the fossil record *at that place* will appear to show a jump.

The gradualists disagreed completely; they maintained that almost all speciation takes place gradually. They accepted the possibility of splitting, as well as drift—after all, the number of species alive today is a lot greater than it used to be, and the new ones must have come from *somewhere*—but saw splitting itself as a gradual separation, not a sudden jump. Eldredge and Gould, in contrast, thought that almost all species arise through rapid changes at splittings, and hardly any changes in species occur by gradual drift. It's a complicated debate, not helped by differences of opinion about what constitutes a species. Cladists, for instance, define species in a manner that precludes drift as a mechanism for species change because to them anything that changes gradually represents the *same* species.

In viewing this debate, mathematicians suggest that the whole discussion is misconceived. Anyone brought up on modern dynamical systems—the best available *general* theory of how systems can change over time—knows that the same system can change either suddenly *or* gradually. The sudden changes are called *bifurcations,* a term that represents a *conceptual* splitting of possible behaviors, rather than an actual splitting of species—and is not confined to splitting into two pieces, despite its usage in everyday language. Imagine a dynamical system with behavior that depends on external parameters—environment, perhaps. Assume that those parameters vary gradually. What does the system do? The obvious answer is that the system changes gradually, too—that continuous changes produce continuous effects.

Obvious—but wrong. *Most of the time,* the effects will change gradually, but every so often, the parameters can hit a critical value at which the change becomes rapid and dramatic. When this happens, we have a bifurcation. For an example, imagine a stick being gradually bent by an external force (e.g., your hands). To begin with, it bends, and the change is just as gradual as the change in the applied force. But then, without anything terribly obvious changing, the stick suddenly snaps. After that, you can continue moving your hands gradually, and the stick (or sticks) again moves gradually with them. There is nothing unusual about this dual behavior: usually smooth, sometimes sudden. It is what nearly all dynamical systems do.

Bifurcations occur when the state of the system changes from being stable to being unstable; the system then seeks a new stable state, which may mean a big change. The gradual behavior occurs when stable states remain stable. Symmetry breaking is a particular type of bifurcation behavior, found in symmetric systems. Even asymmetric systems can, and often do, bifurcate, however.

In the 1960s, Thom introduced some new ideas from pure mathematics into the classification of bifurcations, and Zeeman gave them the name "catastrophe theory" to emphasize the sudden changes involved.[13] Catastrophe theory did not have a big impact on biological modeling—partly as a result of exaggerated criticism in its early days—but it completely revolutionized bifurcation theory. It was a bloodless revolution, accomplished under an assumed name (singularity theory), and it took place largely within mathematics, so hardly anybody noticed. That change in dynamical systems can be sudden was not new, but the possibility of classifying such changes in terms of a sequence of increasingly complex geometric forms *was* new.

The central question, from this point of view, is this: Suppose that a fitness landscape changes gradually, as a result of changes in external parameters. *What should we expect the fitness peaks to do?* You might expect them to move gradually, as well— this seems to be the unspoken assumption behind gradualist thinking about evolution. That expectation is correct if there is only one peak and it does not run into anything else— such as a slope. It is also correct in the especially simple kind of linear mathematics that was common a century ago. But it is *not* true—not even close to truth—for more realistic nonlinear mathematics. The reason is that in nonlinear systems, peaks can be born, can be absorbed, can collide, and can split.

So why not species, too? Admittedly, a fitness landscape is too simple to capture all of the rich reality of biology—but if anything, that reinforces its message: The model shows that both rapid *and* gradual change are natural in any system that occupies the peaks of slowly changing landscapes. There is absolutely no need to choose one or the other as being exclusive, and there is every reason not to. If the *simplest* nonlinear model of transitions in gradually changing fitness landscapes possesses such a rich range of dynamic behavior, surely more complex and biologically more accurate ones should be capable of at least the *same* richness? The gradualist/punctuationist controversy is pointless. Probably, the two schools of thought are both right, some of the time, and both wrong, some of the time. It's time they put their ideas together; neither will win on its own.

So far, the evolutionary mathematics that I've described has been fairly conventional, at least to a mathematician. But some aspects of evolution pose completely new problems for mathematics, and their solution demands the creation of new mathematics. It hasn't arrived yet, but with imagination, we can see it getting nearer. Whether it will reach us gradually or suddenly isn't known.

The need for biologically more realistic models of evolution has stimulated a very different approach to evolutionary modeling, known as the theory of complex adaptive systems, or "complexity theory" for short.[14] Artificial life is a development within complexity theory. Complexity theorists try to model complicated systems of individuals *as* complicated systems of individuals. They don't take shortcuts with average behavior; they don't assume that everything is uniformly mixed. They accept the unique nature of the individual and delight in it. To model evolution, they set up computer models with lots of virtual organisms that obey simple rules of interaction, and then they watch what happens.

Remember the slugs and their selection coefficients? For a complexity-theoretic approach to the same problem, we set up a simulation based on a square grid, say 100 squares by 100. We decide which square corresponds to a piece of bush, or flower, or whatever. Then we populate a randomly chosen selection of these squares with virtual slugs, by assigning a slug *genome* (combination of the alleles under consideration) to each such cell. For example, the twenty-eighth square in row 49 of the grid might be assigned genome red/bush, and so on. Other squares might be virtual predators. Next, we give the computer rules for how these virtual organisms move around the grid and interact with each other. For example, we might decide that at each time step, a slug either moves at random to a neighboring square, or stays put, whereas a predator sees the nearest slug and moves five squares toward it, eating the slug if it reaches the slug's square—meaning that that particular virtual slug is removed from the computer's memory. We set up the rules so that green slugs are less likely to be seen if they are on bushes, rather than on flowers, and so on.

Then we play this mathematical computer game—the technical term is *cellular automaton*—for 10,000 time steps, and we read off the proportions of various surviving slug alleles. In all likelihood, we would run the simulation several hundred times, to ensure that any apparent mathematical patterns are independent of the particular sequence of random events that occurs in a single run. An advantage of the complexity model is that it explicitly incorporates organisms as individuals, rather than by proxy as allele proportions, and it implements natural selection by pitting predators against prey in an environment, rather than simply assigning numerical weights to the probable outcome of such a contest.

Complexity theorists have invented innumerable models in the same spirit: building in simple rules for interactions among many individuals, and then simulating the interactions on a computer to see what happens. The provocative but apt term *artificial life* was coined to describe such activities. A celebrated example is "Tierra," invented by Tom Ray. In Tierra, short segments of computer code compete with each other inside the computer's memory, reproducing and mutating.[15] The source of all

Tierran life-forms is an ancestral organism, a self-replicating segment of computer code occupying 80 bits of memory. In January 1990, Ray released this organism into a primal ocean of random bits in a computer's memory and then left the system to its own devices. Copies of the replicating ancestor quickly took over large regions of memory, but then occasional mutations—computer errors—began to cause changes. New replicating species appeared, some smaller than the ancestor, some bigger. As time passed, the diversity of the ecosystem fluctuated: Sometimes there were very few species, sometimes a lot. It was all rather confusing. Then 45-bit parasites emerged. Lacking their own copying instructions, they borrowed them from nearby organisms. In some runs of the program, the ancestral organism then mutated, becoming 79 bits long and resistant to the parasites, so the parasites died out. In other runs, hyperparasites appeared, which subverted the parasites' method of replication and used this method to replicate themselves. Some of the hyperparasites evolved into social organisms with 61 bits, which replicated only by mutual cooperation. Their existence paved the way for 27-bit cheats, which hijacked the entire program by stealing control from the social organisms.

Tierra may only be a random sea of bits in a computer's memory, but all life is there. It strongly reinforces the view, central to this book, that the patterns exploited by biology arise from mathematics spontaneously, without special effort, prompting, or instruction. Ray did not *instruct* his bit strings to become parasites, or hyperparasites, or to cooperate. They did it anyway.

However, he did instruct them to reproduce, by including an explicit copy command in his computer language. You've got to start somewhere, but the whole story would be a lot more convincing if the ability to replicate itself arose, unaided, because then you'd be modeling the origin of life, not just what happens to it once it has arisen. Ambitious? Improbable? Of course. But science won't advance at all if we pursue only lines of attack for which we can predict success in advance.

In 1996, Andrew Pargellis unveiled his own artificial life program: "Amoeba." Ray had played God by seeding the computer's memory with a specially designed replicator, but Pargellis started with just a random block of computer code. Every 100,000 computational steps, the program wiped out 7 percent of the memory slots and replaced them with randomly chosen commands. He found that about every 50 million steps, a self-replicating segment of code appeared.[16] Replication didn't have to be built into the rules—it just happened.

Systems such as Tierra and Amoeba, without being given any *explicit* instruction to do so, display high-level patterns very similar to those found in real earthly evolution. These patterns include the spontaneous appearance of replicators, spontaneous increases in complexity, rudimentary forms of symbiosis and parasitism, lengthy periods of stasis punctuated by rapid changes—even a kind of sexual reproduction. The message is that all of these puzzling phenomena are entirely natural; they are typical properties of complex adaptive systems. Instead of being surprised when we saw them in the evolutionary record, we should have been surprised if we did not.

These are striking discoveries, but what is their significance? Does artificial life really tell us anything useful about real life? I think it does. I can best describe why by invoking the concept of phase space—a geometric image in which every event that does happen is surrounded by a ghostly halo of nearby events that didn't—but could have. When you set up any mathematical system, be it a classical dynamical system or the kind of thing used by devotees of artificial life, you also implicitly set up a phase space. Phase spaces are *big*—they contain all possibilities, not just a selection. If the rule system is sufficiently rich—which basically means not horribly boring and obvious—then all sorts of possibilities lurk within its phase space. Now we begin to see the significance of mutations in evolution. They don't just make evolution possible; they enable the system to explore its phase space. The states that the system is occupying today may change tomorrow. We also see the role of selection more clearly: It makes the exploration efficient. If all that happened were random mutations, the system would wander around in its phase space like a drunkard, tottering one step forward, two steps back. Indeed the mathematics of random walks shows that such systems spend an awful lot of time revisiting old haunts. With selection, however, bits of phase space that don't work (i.e., don't promote fitness) are eliminated. Selection helps the system to home in on the interesting regions of phase space, the places where useful things happen, the central features of the evolutionary landscape.

The phase space for real earthly evolution is far more complicated than that for Tierra, Amoeba, or even Sommerer and Mignonneau's A-volve. Nonetheless, it plays the same role. Its rules are those of the physical universe. Mutations allow life on Earth to explore the evolutionary phase space, and selection cuts down the possibilities so that evolution doesn't spend all of its time wandering up dead ends or revisiting places that don't contain anything interesting. The combined effect of mutations and selection creates a geography of phase space, making it more like a mountainous landscape than a featureless plain; evolution homes in on the more significant features, behaving *as if* it has goals when actually it is being driven by the geography of its phase space. Evolution does not know where it is heading—but if we could see its phase space, we'd get a pretty good idea.

Stuart Kauffman, a highly innovative scientist who has thought deeply about such matters, sees this kind of structure as being characteristic not just of evolution, but of *any* process that has the ability to complicate and organize itself. Instead of "phase space," he talks of the "space of the adjacent possible." Instead of merely pointing out that phase spaces have a geography, hence a dynamic, he believes that we may soon be able to state precise mathematical laws that govern how a system explores the space of the adjacent possible. He is also convinced that those laws will be a lot closer to "as fast as possible without falling to bits" than they are to "with no purpose and no sense of direction."

6

Flowers for Fibonacci

The beautiful configurations produced by the orderly arrangement of leaves or florets on a stem have long been an object of admiration and curiosity; and not the least curious feature of the case is the limited, even the small number of possible arrangements which we observe and recognise.

D'Arcy Thompson, *On Growth and Form,* Chapter XIV

Nowhere is the second secret of life more evident to a casual observer than in the plant kingdom. There are mathematical patterns in the symmetric arrangement of petals around the edge of a flower, in the way leaves stack above each other along a stem, in the rounded seeds of some plants and the spiky ones of others, in the tiny parachutes that allow still other seeds to float on the breeze. Even the irregular geometry of trees hints at some more elusive pattern, some kind of self-referential geometry in which small pieces of a tree bear an uncanny resemblance to the whole—so much so that model makers use twigs to represent scaled-down trees. The plant world has borrowed such structures from physics and still employs them in much their original form.

Other features of plants, however, have evolved farther away from raw physics. Their mathematics, if they have any, has been submerged beneath layer upon layer of evolutionary tinkering. There are flowers that mimic female flies in order to fool the male into intimate contact that spreads pollen from plant to plant. Such plants have interacted with the evolution of insects for hundreds of millions of years—no free-running mathematics there. Any understanding of the forces that shape plants must distinguish those presented free of charge by physics from those that have been molded by evolution; it must disentangle universal mathematical constraints from flexible genetic instructions.

The mathematical aspects of plants have long been recognized. D'Arcy Thompson clearly saw that the strange numerology of the plant world has important implications for the biology of plant development. Thanks to contemporary work in dynamics, we now have a rather clear idea of what is involved in such biology. Following a well-established tradition, which he traced back to Leonardo da Vinci but speculated might well go back to the ancient Egyptians, Thompson observed that the plant kingdom has a curious preference for particular numbers and for spiral geometries, and that the numbers and the geometries are closely connected. We have already observed that the numbers that arise in plants—petals, sepals, all sorts of other features—are usually taken from the Fibonacci sequence 1, 2, 3, 5, 8, 13, 21, 34, 55, 89, in which each number is the sum of the two that precede it. Most exceptions either involve (a) the doubling of these numbers, a trick that is made possible by certain peculiarities of plant chromosomes but that still uses the Fibonacci pattern; or (b) the so-called "anomalous sequence" 1, 3, 4, 7, 11, 18, 29, which exhibits the same additive pattern as the Fibonacci sequence, but starting with different numbers.

The study of geometric and numerical patterns in plants is known as "phyllotaxis," and it has a *huge* history and literature. The spirals of the fir cone were studied in the mid–eighteenth century by two mathematicians, Charles Bonnet and G. L. Calandrini. Around 1837, the pioneering crystallographer Auguste Bravais and his brother Louis made a major contribution to phyllotaxis theory, when they discovered the most significant single regularity in a growing plant—a particular angle, described later in this chapter, that is universal to plant geometry. In 1872, the Scottish mathematician Peter Guthrie Tait, an incorrigible dabbler in mathematical curiosities, used the geometry of lattices—patterns of repeating parallelograms—to show that whenever one particular system of spirals occurs, other associated spirals must also become evident to the eye. This is why we notice *two* families of spirals formed by adjacent seeds, both related by adjacent Fibonacci numbers, and both very different from the true source of pattern—as we shall see. All of these approaches were purely descriptive—they didn't explain *how* the numbers related to plant growth; they just sorted out the geometry of the arrangements.

The search for an explanation of the Fibonacci numerology of plants has been going on for more than three centuries. Finally, it seems that the goal has been achieved: In 1992, two French mathematicians, Yves Couder and Adrien Douady, set the seal on centuries of research by tracing Fibonacci numerology to natural dynamical constraints on plant development. Their work, the main focus of this chapter, demonstrates conclusively that the apparent mathematical patterns in plants do indeed arise from universal laws of the physical world. They are not merely genetic accidents reinforced by evolution. As always, however, the physical laws must work hand in hand with the plant's genes, for without genes, there would be no plant to develop in the first place.

By a curious quirk of history—or maybe yet another layer of depth in the mathematical nature of physics—Fibonacci's rabbit puzzle contains further messages for the mathematics of plants. It is not just the numbers that matter; it is also the manner in which they arise. Fibonacci numbers are the visible tip of a marvelous mathematical theory of branching structures—the rabbits' family tree, so to speak, and I'm not talking metaphorically. The same scheme illuminates not just the numerology of plants, but also their entire form—the way they, too, branch. We can now—in computer simulation—*grow* realistic grasses, flowers, bushes, and trees from mathematical rules, and there is a strong suspicion that those rules lie at the heart of how plants themselves grow.

Numerology first. It turns out that the best way to understand the occurrence of Fibonacci numbers in plants is *not* to focus on their arithmetic. In a sense, the additive pattern of the numbers is a coincidence, a mathematical consequence of Fibonacci numerology but not a significant part of its cause. The best way into the problem is to look at plant geometry.

A good starting point is one of Thompson's main examples of patterns in plants: the arrangement of sunflower seeds (Plate 6). The mathematical pattern here is very striking. The seeds occur in two families of spirals— one winding clockwise, the other counterclockwise, and appearing to fit through each other. In this example, there are 34 clockwise spirals, like the spokes of a wheel, but curved, and 55 counterclockwise spirals. The two numbers are different, but both are Fibonacci numbers—and consecutive ones in the series. The precise numbers depend on the species of sunflower, but you often get 34 and 55, or 55 and 89, or even 89 and 144. Daisy heads have similar patterns in them, but they are smaller. Pineapples have 8 rows of scales—those diamond-shaped bits—sloping to the left, and 13 sloping to the right. Cones of the Norway spruce have 5 rows of scales in one direction and 3 in the other; the common larch, another conifer, has 8 and 5; the American larch 5 and 3 again. If genetics can choose to give a flower any number of petals it likes, or a pine cone any number of scales that it likes, why such a preponderance of Fibonacci numbers? It seems more likely that the numbers arise through some more mathematical mechanism, such as a dynamic constraint on plant development, and the evidence supports this view.

Plants begin their existence as seeds, which germinate to put out many tiny roots, which grow downward into the ground, and a single shoot, which grows upward into the sunlight. All of the curious numerology of the adult plant is already present, albeit invisible to the unaided eye, in that early shoot. It is created by dynamic conditions at the very tip of the shoot, where new cells are appearing as the plant grows. As the new cells jostle into position, they are already lining up into Fibonacci spirals. If you look at the tip of the shoot under a microscope, you can see the bits and pieces from which all the main features develop—leaves, petals, sepals, florets, or whatever. At the center of the tip is a circular region of tissue called the "apex"; around the apex, one by one, tiny lumps form, called "primordia." Each primordium migrates away from the apex—more accurately, the apex grows away from the lump, leaving it behind—and eventually, the lump develops into a leaf, petal, or the like. Moreover, the general arrangement of those features is laid down right at the start, as the primordia form. So the heart of the problem is to explain why you see spiral shapes and Fibonacci numbers in the primordia, as all the varied Fibonacci features of plants are simple consequences of this basic geometric structure.

The first step is to appreciate that the spirals that are most apparent to the human eye—known to botanists as "parastichies"—do not provide a direct representation of the actual pattern of growth of the plant's tip. In a sense, they are just optical illusions. The most important spiral is formed by considering the primordia *in the order in which they appear* (Figure 46). Primordia that pop into existence earlier will migrate farther, so you can deduce the order of appearance from the distance away from

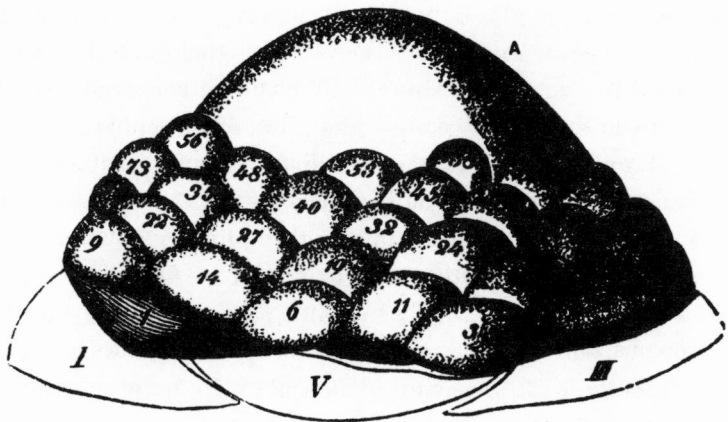

Figure 46
Primordia numbered in order of appearance.

the apex. It turns out (Figure 47) that the primordia are spaced rather sparsely along a very tightly wound spiral, called the "generative spiral." The human eye picks out the parastichies because they are formed from neighboring primordia, but it is the generative spiral that really sets up the mathematical patterns.

The Bravais brothers' great contribution was to find the mathematical rule that underlies the spacing of primordia along the generative spiral.[1]

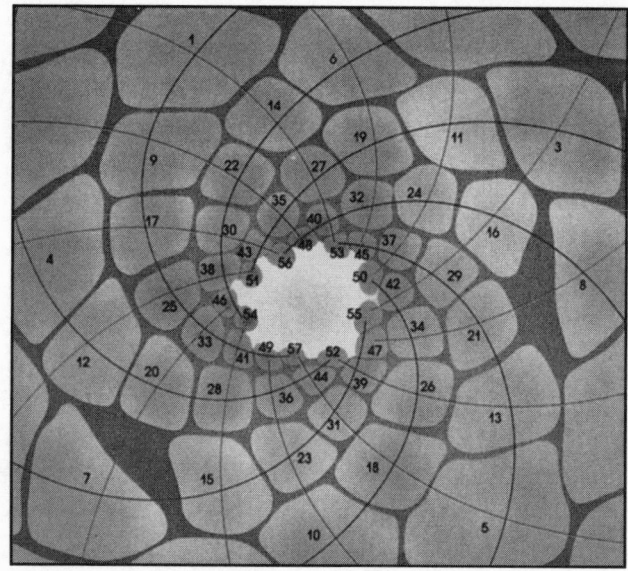

Figure 47
Spacing of primordia, here shown in a cross section of a bud.

They measured the angles between successive primordia, as seen from the center of the apex. Their first discovery was that the angles are pretty much equal; their common value is called the "divergence angle." (For example, look at the angles between the primordia numbered 29 and 30, or 30 and 31, and so on.) Their second discovery pinned the divergence angle down completely: It is usually very close to 137½ degrees. To appreciate the mathematical significance of this number, take consecutive numbers in the Fibonacci series, such as 34 and 55; form the corresponding fraction 34/55, and multiply by 360°. The result is approximately 222.5°. Now, we can measure angles externally or internally, and because 222.5° is more than 180°, we should subtract it from 360°, which yields that mysterious 137.5°.

It can be proved that the ratio of successive Fibonacci numbers, as they get bigger and bigger, approaches closer and closer to the value 0.618034. For instance, 34/55 = 0.6182, already quite close. The precise limiting value is $(\sqrt{5} - 1)/2$, the "golden number,"[2] often denoted by the Greek letter phi (ϕ). Accordingly, we call 137.5° the *golden angle*. A more precise value is 137.50776°.

The Bravais brothers discovered that the golden angle is the one favored by nature, but they did not find out why it is. It is certainly a very curious number for nature to employ, so there has to be a good reason. The first glimmer of insight into why the golden angle is a consequence of sensible physical principles—natural, rather than accidental—was gained by G. Van Iterson in 1907. He worked out what arrangement you get by plotting successive points on a tightly wound spiral at angles of 137.5°. He showed that because of the way neighboring points align, you get two families of interpenetrating spirals—one winding clockwise and the other counterclockwise (Figure 48). Because of the close relation between Fibonacci numbers and the golden number, the numbers of spirals in the two families are consecutive Fibonacci numbers—*which* Fibonacci numbers depends on the how tightly the spiral is wound.

Van Iterson's theory was again purely descriptive: It exhibited the key patterns in the geometry of the arrangement but did not explain their physical or biological causes. The next advance, made by H. Vogel[3] in 1979, got a lot closer to pinning down the *cause* of the patterns. He represented the seeds in the head of a sunflower by equal circular discs, and he worked out what spacing rule—assuming a constant divergence angle—would pack those discs together as closely as possible. His computer experiments showed that if the divergence angle is *less* than 137.5°,

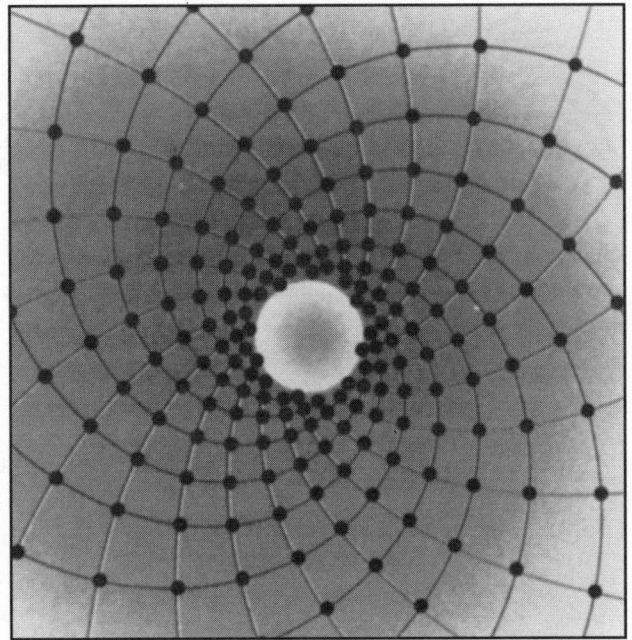

Figure 48
Circular discs spaced along a tightly wound spiral, separated by the golden angle, create the optical illusion of Fibonacci spirals.

then gaps appear in the seed head, and you see only one family of spirals. Similarly, if the divergence angle is *more* than 137.5°, then gaps again appear in the seed head, but this time you see only the other family of spirals. The golden angle thus turns out to be the *only* angle at which the seeds pack without gaps; and when they do, both families of spirals occur simultaneously (Figure 49). In short: The most efficient packing, the one that makes the most solid and robust seed head, occurs when the divergence angle is equal to the golden angle.[4] *That's* what makes the golden angle special, and it all stems from the geometric principle of efficient packing.

Biologists are advisedly wary of explanations of form that point to some elegant or desirable property of a system without explaining how that elegant property can be achieved. For example, "birds have wings in order to fly" is not an explanation of wings because it doesn't tell us *how* wings evolved, or even whether they *can* evolve. It just says they would be a great idea if they could evolve. Evolution doesn't select some feature because it would be a good idea; there has to be a sensible way to get to it. Efficient packing is a neat idea, and it obviously helps to produce a strong, solid plant. However, the desirability of efficient packing does not guarantee the existence of a mechanism to produce it.

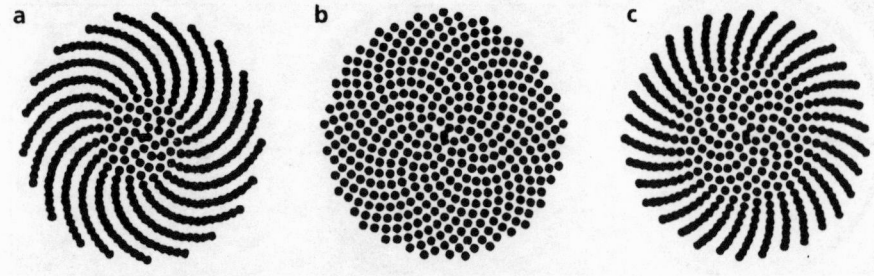

Figure 49
Packings created by divergence angles of (a) 137.3°, (b) 137.5° (the golden angle), and (c) 137.6°. Only in case (b) is the packing efficient.

It was just such a mechanism that Douady and Couder discovered.[5] They obtained the golden angle as a *consequence* of sensible dynamics, rather than by postulating it directly on grounds of efficient packing. Their starting point was the assumption that successive elements of some kind—representing primordia—form at equally spaced intervals of time somewhere on the rim of a small circle, representing the apex; and that these elements then move directly outward. They also assumed that the elements repel each other, like equal electrical charges or magnets with the same polarity. This repulsion ensures that the outward motion keeps going, and that each new element appears as far as possible from its immediate predecessors. In other words, each new primordium pops up in the biggest gap.

It's a good bet that such a system will satisfy Vogel's criterion of efficient packing because if you keep filling the biggest gap, then only small gaps will remain. Therefore, you'd expect the golden angle to show up of its own accord—and so it does, as Douady and Couder demonstrated, using a physical analogue: a circular dish filled with silicone oil and placed in a magnetic field. They let tiny drops of magnetic fluid fall at regular intervals into the center of the dish. The drops were polarized by the magnetic field, and they repelled each other; they were given a boost in the radial direction by making the magnetic field stronger at the edge of the dish than it was in the middle. The patterns that appeared (Figure 50) depended on how big the intervals between drops were, but a very prevalent pattern was one in which successive drops lay on a spiral with a divergence angle very close to 137.5°, the golden angle, giving a sunflower seed pattern of interlaced spirals. Douady and Couder also carried out computer calculations, with very similar results.

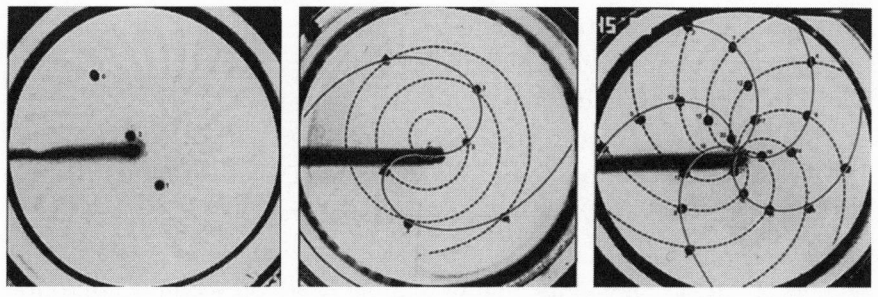

Figure 50
Fibonacci spirals observed in an experiment with electrically charged oil drops.

To clinch the matter, in the mid-1990s, a complete mathematical proof of their picture of the dynamics was given by M. Kunz.[6] The two French mathematicians found that the divergence angle depends on the interval between drops, according to a strikingly geometric bifurcation diagram— a branching pattern of wiggly curves (Figure 51). Each section of a curve between successive wiggles corresponds to a particular pair of numbers of spirals. The main branch runs along very close to a divergence angle of 137.5°, and along that branch, you find spiral arrangements using all pos-

Figure 51
Bifurcation diagram for spiral numerology. The vertical axis, labeled "G," corresponds to the rate at which new primordia appear, relative to the speed with which they move away from the tip of the growing shoot.

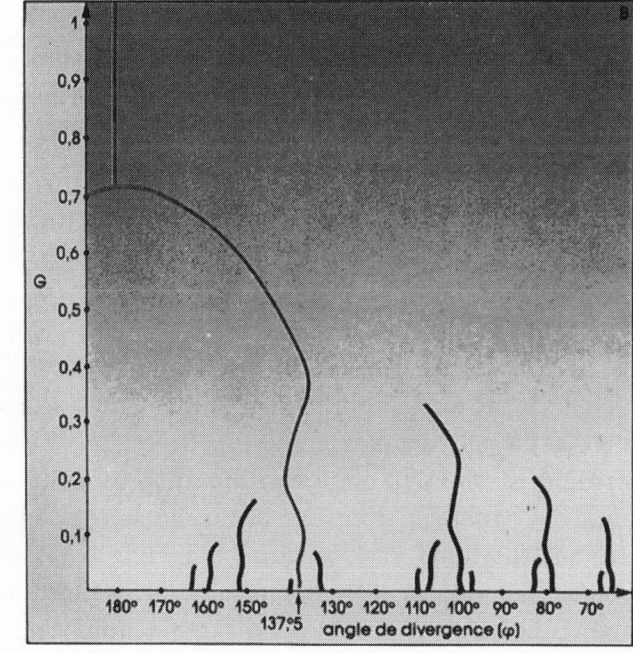

sible pairs of consecutive Fibonacci numbers, one after the other in numerical sequence. The gaps between branches represent bifurcations where the dynamics undergoes significant changes.

However, there are also branches in the diagram that *aren't* close to 137.5°. This turns out not to be a defect of the Douady–Couder theory, but a strength. Strikingly, the main exceptional branch corresponds to the anomalous series 3, 4, 7, 11, 18 . . . , which is also the main exception to Fibonacci numerology in the plant kingdom itself. So with the appropriate timing for the drops, the same model produces the main exceptions to the Fibonacci rule, as well as the Fibonacci rule itself—which shows why the exceptions occur, while making it clear that they aren't really exceptions at all. Of course, nobody is suggesting that botany is *quite* as perfectly mathematical as this model. In particular, in many plants, the rate of appearance of primordia can speed up or slow down, and changes in morphology—whether a given primordium becomes a leaf or a petal, say—often accompany such variations. Could the role of plant genes be to affect the *times* when primordia pop into being? Certainly the genes don't need to tell primordia how to space themselves out—that's done by the dynamics. Plant growth is a partnership of physics and genetics, and you need both to understand it.

With this final piece of the jigsaw in place, the striking numerical patterns in plants are revealed as an inevitable consequence of the dynamics of plant growth and are thus governed by strict mathematical laws. If, on a distant world, with an entirely alien kind of life—one with genetics not based on DNA—organisms like plants arose, growing in the same general way, then they, too, would exhibit Fibonacci numerology. Fibonacci numbers are not an accident; they are consequences of universal geometry, the crystallography of plant structures. Indeed, plants can no more avoid Fibonacci numbers than salt crystals can avoid being cubical.

At the same time that Douady and Couder were unraveling the dynamics of primordia, other mathematicians, notably Aristid Lindenmayer, were trying to explain the *shapes* of plants. In particular, they wanted to explain the patterns in which plants branch. If you look closely at plants and shrubs, you will often find that their branches don't divide at random, but have regularities. Perhaps branching will lead to one long stem and one short one, each of which branch in turn according to the same pattern, for instance. By a curious quirk of fortune, it turns out that branching patterns depend on mathematics that also goes back to

Fibonacci's rabbit problem. Now, however, it is the rabbits' family tree, not the size of their population, that matters.

The new concept required to determine the branching rules of plants is that of a fractal. A *fractal* is a geometric shape that has intricate structure on all scales of magnification. Fractals are probably one of the best known new mathematical concepts, probably because of their intriguing beauty. You can buy fractal T-shirts, beer mugs, even jigsaw puzzles. The name was coined by Benoit Mandelbrot, who pioneered modern fractal geometry and many of its applications. Fractals are a new way to model the irregularities of nature—things such as clouds, mountains, river deltas, cratered lunar landscapes, the bays and promontories of continents— and trees, leaves, grasses, ferns. . . . In a sense, fractals seem to have captured a hitherto unformalized feature of the geometry of nature, one that has a natural appeal to our aesthetic sense. Maybe that's why we find them on so many T-shirts, beer mugs, jigsaw puzzles. . . .

Fractals reveal hidden regularities in nature. The simplest fractals are *self-similar,* which means that their shape is assembled from copies of itself. When a tree trunk first splits into two branches, each branch looks very much like a complete tree. Sometimes you can even plant them in the ground, and a new tree will grow. In this sense, a tree is made from two copies of itself. They may not be perfect copies—a tree is *statistically* self-similar, rather than rigidly so—but each piece looks like a tree. We don't expect to find perfect fractals in the real world, any more than we expect to find perfect spheres. Both are mathematical ideals; real-world versions will be imperfect. Real-world fractals don't have intricate structure on *any* scale of magnification—just on a wide range of scales. Look *really* closely at a tree, and you see atoms, not tiny trees. The plant world is full of forms that can reasonably be modeled by fractals. A striking example is cauliflower (Plate 7), the florets of which are themselves composed of florets which are composed of . . . well, you get the idea. It doesn't go on forever, like a mathematician's fractal, but it does go on for several stages more than you might expect.

Lindenmayer developed an algebraic system for generating fractal forms similar to the branching structure of plants, a kind of "branching grammar." His ideas[7] are now known as "Lindenmayer systems," or "L-systems" for short. The simplest L-system goes back to the original source of Fibonacci numbers, a puzzle about rabbits. Remember how the problem goes: Assume that at breeding season zero, we start with precisely one pair of immature rabbits, which mature for one season; every season, each mature pair produces just one immature pair, which in turn takes one season to mature; rabbits are immortal. How many pairs of rabbits will there be in each season?

The orthodox way to solve this puzzle is to set up a system of equations and to solve them using algebra. Here, however, we look at the branching rules for the rabbit family tree because that approach will lead us to branching rules for plants. Introduce the symbols I for an immature pair, M for a mature pair. The branching rules take the form

I → M   (an immature pair becomes mature one season later)
M → MI   (a mature pair remains alive and also generates one immature pair)

We start with one immature pair (I) and repeatedly apply the two branching rules to get the sequence of symbol strings

I → M → MI → MIM → MIMMI → MIMMIMIM → . . .

At each step, we apply the branching rules to every symbol in the string, and this leads to the next string. Now we just have to count the symbols. For example, the final generation listed contains five M's and three I's, a total of eight pairs of rabbits. The numbers 3, 5, and 8 are consecutive Fibonacci numbers, and the pattern continues indefinitely.

Instead of counting the symbols, we can *interpret* them as actual branches in a tree diagram (Figure 52). Now we are modeling a plant that

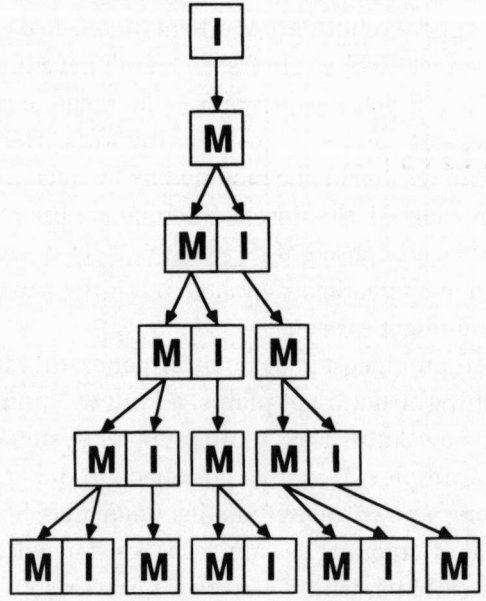

Figure 52
Geometric representation of the L-system (Lindenmayer system) for the family tree of Fibonacci's rabbits.

has two cell-types: immature ones, which mature for a season and then branch, and mature ones, which produce an immature side-branch while continuing to grow themselves. No longer a family tree of rabbits, this is just a tree—or some treelike plant. Lindenmayer developed other grammars of the same general kind; some can be interpreted as a flexible kind of cellular automaton. For example, Figure 53 shows the fifteenth growth stage of a model of the red alga *Callithamnion roseum,* based on the following rules:

◇ At the base of the main filament are three cells that do not have branches.

◇ Each successive cell above these has one branch.

◇ In all stages, four cells below the tip of the main filament have no branches.

◇ Each branch repeats the pattern of the main filament.

Various sample L-system plants are shown in Figure 54. There are probabilistic L-systems, as well as deterministic ones, and also context-sensitive L-systems, the grammatical rules of which depend on contextual constraints. Context-sensitive L-systems can model real plants very convincingly (Figure 55). Przemyslaw Prusinkiewicz has wedded L-systems

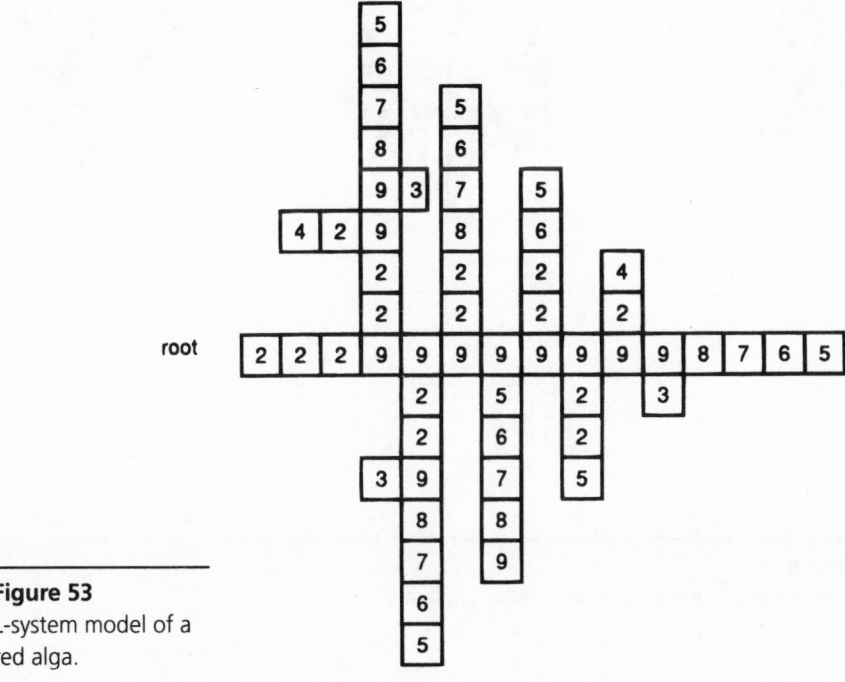

Figure 53
L-system model of a
red alga.

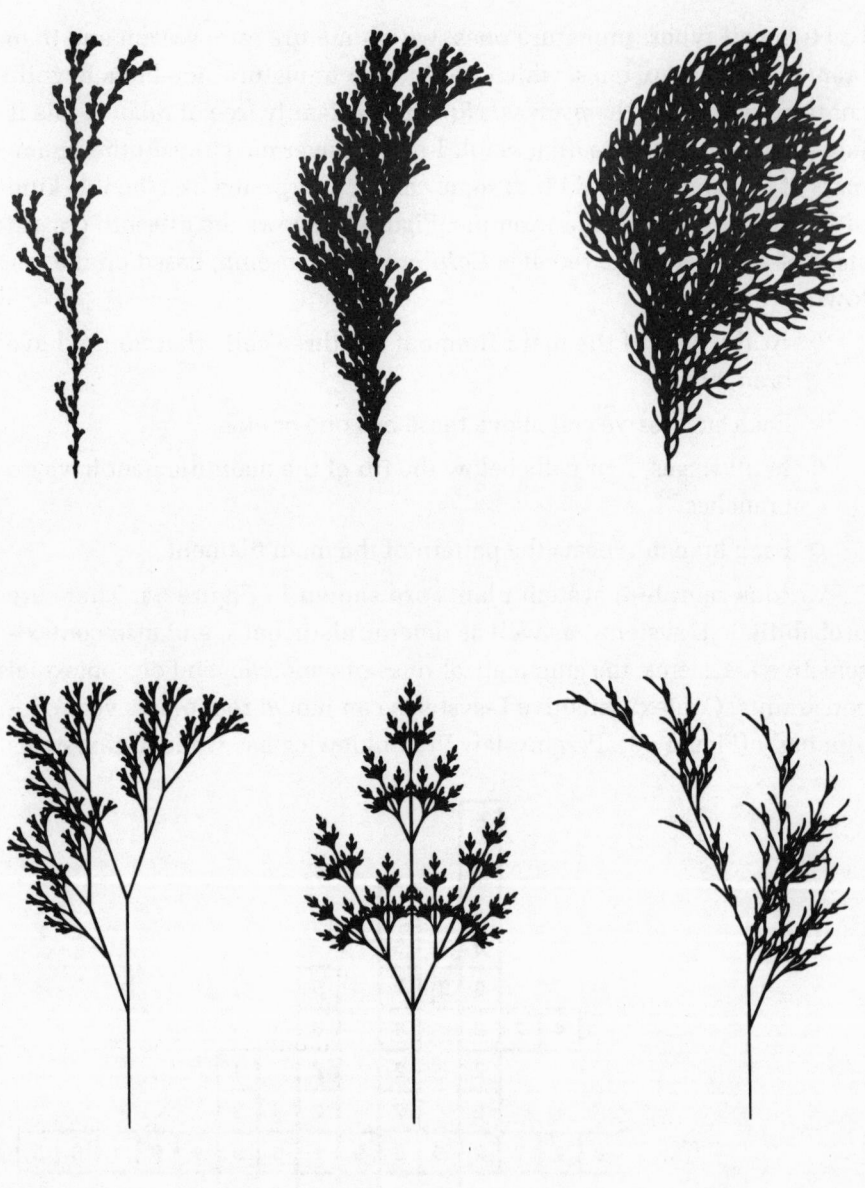

Figure 54
Plant shapes created by deterministic L-systems.

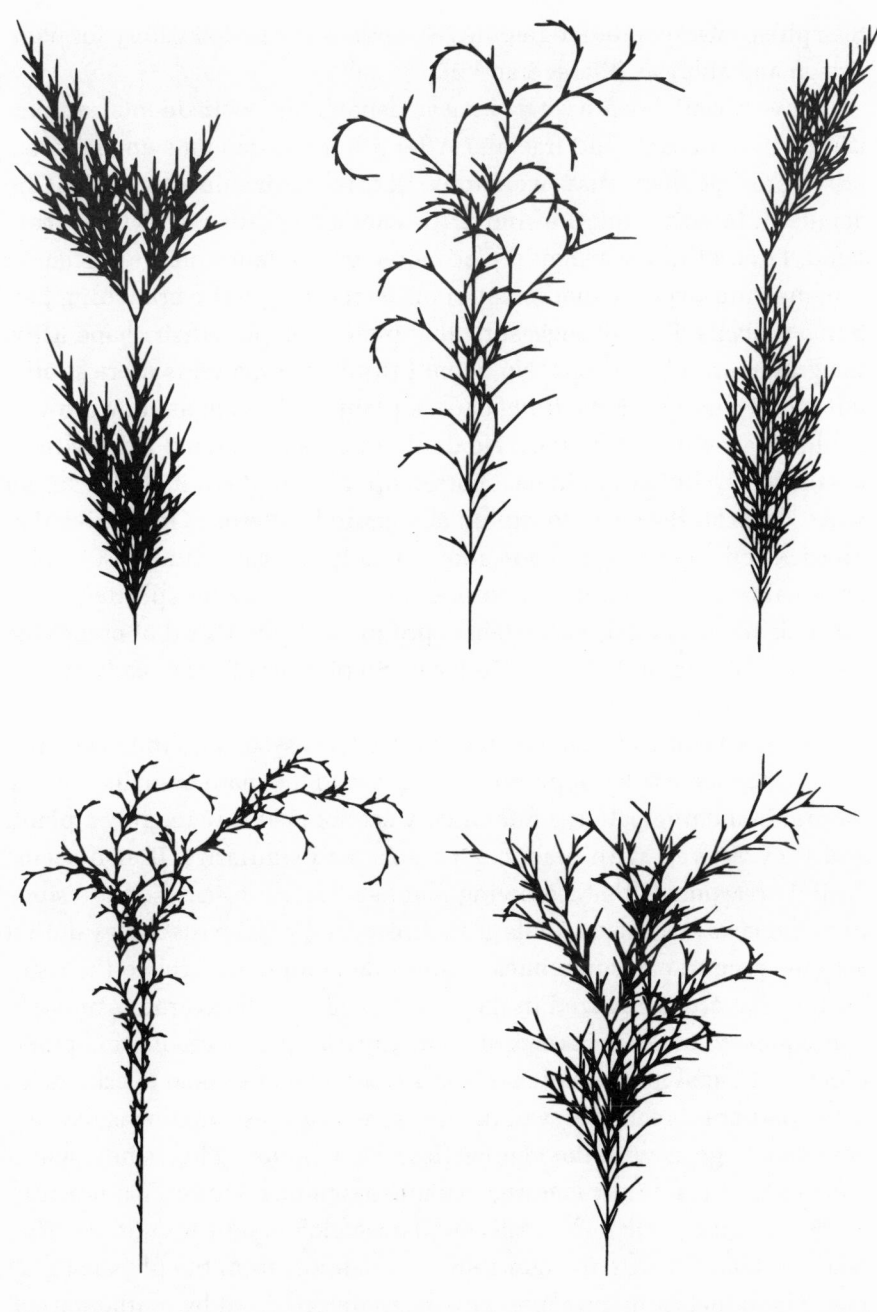

Figure 55
Plant shapes created by context-sensitive L-systems.

to sophisticated computer graphics to create a virtual laboratory for plant design and analysis (Plates 8 and 9).

All very well, but are we looking at visual puns—coincidental similarities between plants and fractals? After all, if you generate enough fractals, some of them may well look like recognizable plants, just by accident. I don't think the similarities are a coincidence. On the other hand, I don't think we should place too much reliance on the particular scheme employed by mathematicians to reproduce the branching patterns of plants. I'm not suggesting that plants compute their shape using the grammar of L-systems. No, what I think is going on is more subtle, but more sensible. *Something* causes plants to branch as they grow. It could be genetic instructions: No doubt those are what tell a daisy to be a daisy and a buttercup to be a buttercup. Nonetheless, it would surely make sense for the genes to exploit any natural patterns of branching that physics and chemistry provide automatically, by way of universal mathematical rules. Branching is a form of bifurcation; the tip-splitting instability of an ice crystal, for instance, produces its own kind of branching pattern, not unlike that found in ferns. So physics offers a ready source of branching processes.

The key feature of Lindenmayer's rules for L-systems is their self-similarity: The same rules apply to the twigs as to the main branches. These natural branching patterns offer a very economical way to grow a plant, and they fit well with plants' geometric self-similarity. It also seems entirely reasonable that a growing plant should have access to the *same* physical laws at all stages of its growth, especially given that twigs stuffed into the ground can sometimes sprout into complete trees. So the regularities that are formalized in the grammar of L-systems are presumably consequences of the dynamics of plant growth—just as Fibonacci numerology is. Fractal mathematics is the way we humans come to grips with these particular regularities of the universe —just as calculus is the way we come to grips with the regularities of gravitation. This is how mathematics works: It is *our* language for understanding nature's patterns.

What of the genes? They tell the plant which plant it should be. That is, they select a particular assortment of patterns from the physical grab bag. The patterns themselves, however, are provided by mathematical rules—and unless we understand those rules, and what mathematics they conceal, the true role of the genes will forever remain mysterious.

7

Morphogens and Mona Lisas

At the same time, in attempting to explain the
phenomena . . . all that we are meanwhile justified
in doing is to try to shew that such and such actions
lie *within the range* of known physical actions and
phenomena . . . somehow or other capable of being
referred to dynamical laws, and to the general principles
of physical science.

D'Arcy Thompson, *On Growth and Form,* Chapter IV

At the edge of the woodlands, the sunlight can still penetrate the barrier of trees, making dappled patterns on the ground, rippling in glowing patterns as the trees sway gently in the summer breeze. . . . A butterfly flits from flower to flower, gliding lazily on fragile, multicolored wings. The flower has five petals, a Fibonacci number. *We* know why, but the butterfly's rudimentary mind is firmly focused on the nectar that the plant contains.

As we watch, it slowly dawns on us that there is mathematics, too, in butterflies. Butterfly wings are a glorious riot, not just of color, but of geometry. There are the peacock butterfly's deep blue spots, arranged in pairs, one on each wing, glowing like fox eyes in the headlights of a passing car. Or the rows of rectangular yellow patches on the swallowtail, divided by tidy black lines like a child's notebook, spaced neatly for tiny notes. The zebralike wingtips of the painted lady, white splotches on a background of black velvet . . . the three rows of black and orange notches on the leading edges of the small tortoiseshell . . . the intricate veins of the monarch, spiky black lines radiating in elegant curves across its dusky brown wings . . .

And what of larger animals? Who can remain unmoved by the striking beauty of the tiger—what Blake called its "fearful symmetry"? Black stripes on orange-brown body, packed together tightly over the face, picked out around the eyes and jaws in white, merging flawlessly into four neat, parallel rows of whiskers (Plate 10)—or the leopard, with its proverbially indelible spots—each spot a patch of brown flanked by black, separated by corridors of white (Plate 11). The clouded leopard, with large gray and black patches on a buff body, so prized for its fur that it is an endangered species, still hunted even though it is protected by law. The common genet, with its striped tail and spotted body—part leopard, part (so it seems) zebra. Theirs is a subtle geometry—but geometry nonetheless.

In the animal kingdom, we don't find the same striking *numerical* patterns that we do in plants. Except, perhaps that most animals seem to have an even number of legs. Nonetheless, we most surely do find patterns: patterns of shape, and patterns in surface markings.

The first person to apply mathematics to the markings of animals was Alan Turing, a mathematical logician best known for his pioneering role in the invention of the computer and his deep understanding of the inherent limits of computational mathematics. Turing discovered an unexpected unity in animal markings: They can all be produced by the same type of equation. Such equations describe what happens to chemicals when they react together and diffuse over a surface or through a solid medium, so we call them "reaction-diffusion equations." Turing's equations do not match biology precisely; they are best viewed as a particularly simple example of the *kind* of mathematical scheme that must govern pattern formation in animals.

Also, sometimes, the equations do match biology, at least in important respects. For example, there is a beautiful tropical angelfish, *Pomacanthus imperator*. Like all angelfish, it has a flat body, wide at the back, pinching down to a beaklike snout at the front. It is smartly marked, with a white face and a black stripe across the eyes, thinly bordered in deep

Plate 1 Fir waves in a New Hampshire forest form moving crescents of dying trees.

Plate 2 Patterns in the BZ reaction.

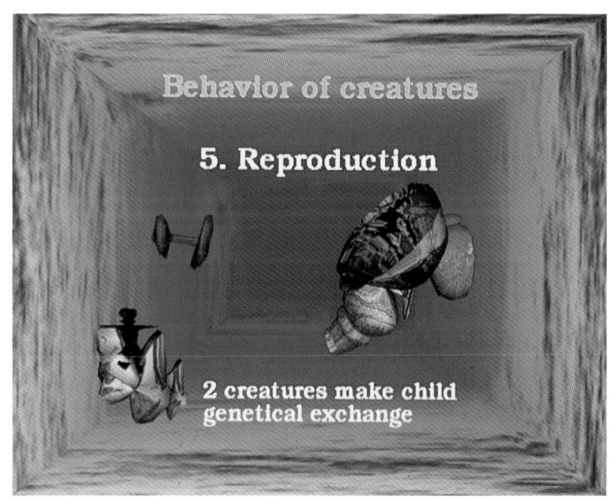

Plate 5 Sommerer and Mignonneau's artificial life exhibit "A-volve."

Plates 3 and 4 follow.

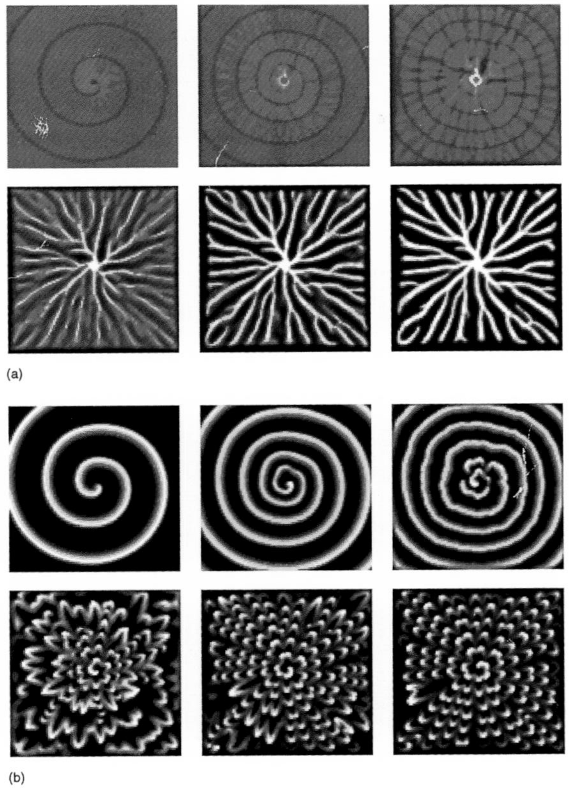

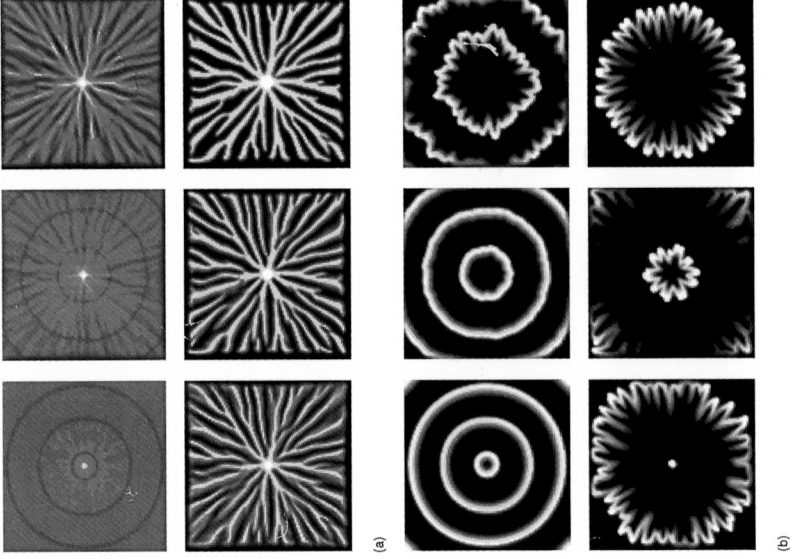

Plate 3 Mathematical model of slime mould aggregation showing spirals.

Plate 4 Model of slime mould aggregation showing concentric rings.

Plate 6 Spiral seed patterns in the head of a sunflower.

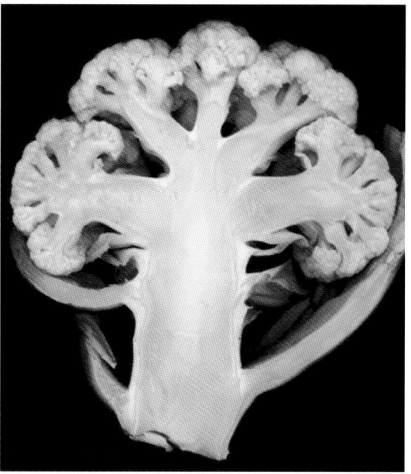

Plate 7 A cross section of cauliflower shows its fractal shape.

Plate 8 A computer-generated image of a wild carrot plant.

Plate 9 A computer-generated image of a mint plant.

Plate 10 The vivid stripes of a tiger.

Plate 11 The distinctive spots of a leopard.

Plate 12 A real Amoria elhoti shell at left and a computer simulation at right.

Plate 13 A computer generated image of a volute shell.

Plate 14 An *Acetabulatia acetabulum* colony, more popularly known as mermaid's cap, in its natural habitat.

Plate 15 The angelfish *Pomacanthus semicirculatus*.

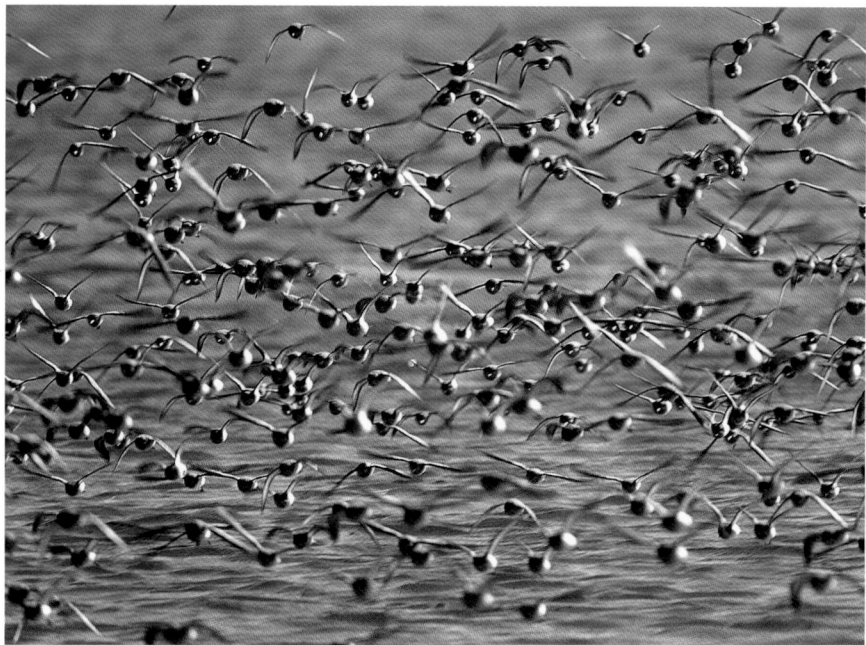

Plate 16 A flock of Western Sandpipers in flight over the ocean.

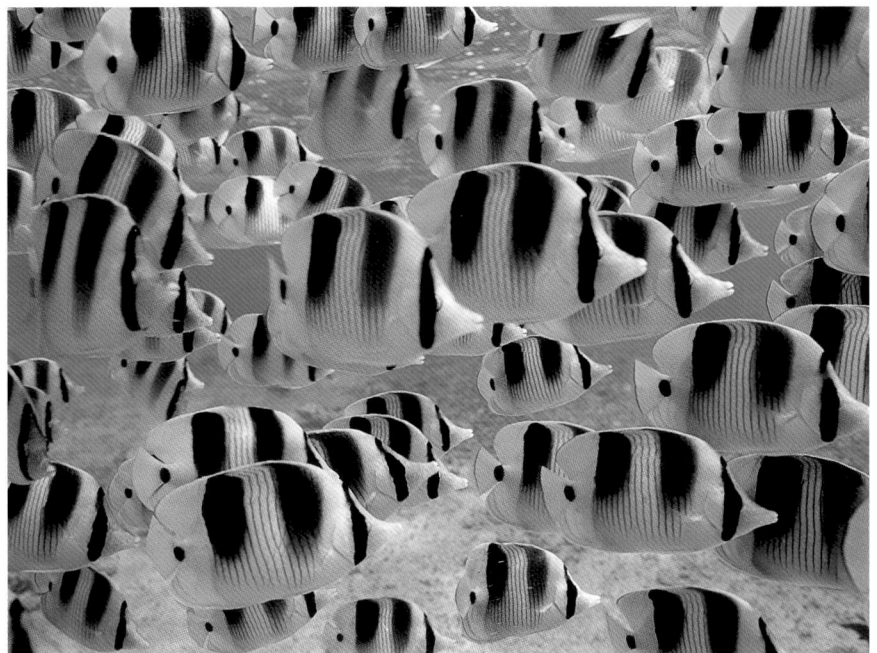
Plate 17 A school of Pacific double-saddle butterfly fish in Bora Bora.

Plate 18 The web of an orb weaver spider.

Plate 19 An orchid fractal, related to crowd flow models.

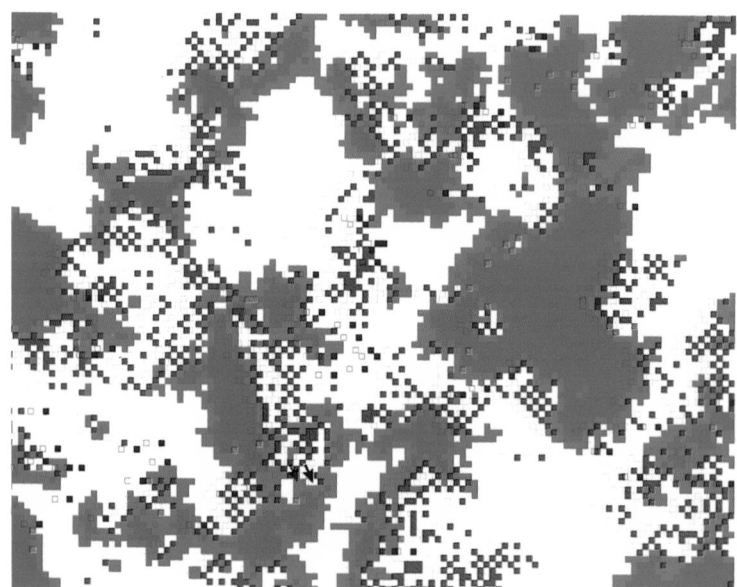

Plate 20 A cellular automaton model of part of the ecosystem of the Persian Gulf. Green corresponds to kelp; purple to sea urchins; red to lobsters; blue to lobsters that have not been eaten for at least four days; and white to bare rock.

blue. Along two thirds of its body run parallel stripes of yellow and purple. Stripes are perhaps the archetypal Turing pattern, but there are difficulties in matching the mathematics to the biology. Some Turing patterns move. If the stripes of the angelfish were to be explained by Turing's theories, then those stripes would have to move, too.

They do. In 1995, two Japanese scientists, Shigeru Kondo and Rihito Asai, discovered that over a period of several months, the stripes on the angelfish migrate across its surface. As they do so, certain defects in what would otherwise be a pattern of regularly spaced stripes, known to physicists as "dislocations," break up and re-form in characteristic ways: exactly the ways that Turing's equations predict.

Turing's theory has scored other striking successes, a very impressive case being the markings on seashells.[1] Shells grow along their edge, where the animal's mantle secretes new structural material—and pigment. The pigment is what gives rise to patterns—sometimes trite, sometimes elaborate. In buttontop shells, Asian mud-dwellers, the spiral form of the shell merges into a flat cone, but each twist is marked with stripes of brown, red, or green. The turban shell from Australia is aptly named, wound in alternating bands of green and stripy white and brown. Sundials, found in the Atlantic, are flattened coils with regular dark dots on a pale background. The wentletraps of Europe, the Caribbean, and Japan are elongated, decorated with strange flanges, sometimes looking like basket weave. Conches are covered in irregular, wavy lines of brown and white, like brushmarks. Moon snails bear a variety of patterns: The china moon displays rows of neat black squares; the zebra moon has parallel wavy stripes; the butterfly moon has stripes that are themselves striped; the maculated moon bears a tidily spaced array of tiny triangles. Cowries are mostly spotty—big spots, small spots, regular spots, random spots. Bonnet shells look like mint humbugs, brown stripes alternating with white. The stripes can run from the tip of the shell to its mouth, or around and around its body. Sometimes, the stripes break up into a series of squares, like this: ■ ■ ■ ■. Particularly enigmatic are the olive shells: For example *Oliva (O.) porphyria* has a rounded, elongated shape, decorated with apparently random brown triangles on a pale background. Similar patterns can be found on baler shells, volutes, and above all cone shells. The living textile cone has a light brown background, along which run wavy dark brown stripes; overlaid like appliqué patches are white triangles, each having a thin, black border, looking like sharks' teeth. The glory-of-the-seas is olive-colored, decorated with a wide spiral of pale brown, almost invisible behind an intricate weave of diminutive spindly triangles. Indeed, the cone shells are perhaps the most versatile of all—their surface markings resemble cloth, marble, mountain ranges, old computer-punched tapes, lines of Morse code, and random splotches deposited by a careless housepainter.

The sheer diversity is astonishing, the beauty breathtaking. It seems hardly possible that such natural beauty can be generated by a mathematical formula. Yet it can. In fact, it is the same formula that Turing introduced to model animal markings. Even more surprising is the fact that not only *can* Turing models create such irregular patterns, but they manage it rather easily. For example, Plate 12, created by Przemyslaw Prusinkiewicz and Deborah Fowler, shows a real Amoria elhoti shell and a computer simulation with the appropriate pattern superimposed. Hans Meinhardt has made a wonderful study of the markings on shells, showing that despite their great variety and intricacy, they can all be produced by Turing's equations.[2] Plate 13 shows another example of Meinhardt's work. What's more, D. M. Raup, C. Illert, and others have even written down equations for the *shapes* of shells.[3]

Despite these remarkable achievements, I don't want you to think that Turing's theory has been a complete success. His equations also have their problems. They don't always agree with experiments, especially experiments that grow organisms at different temperatures from usual. Also, some instances of patterns in animal markings seem to arise for very different reasons, such as rows of cells turning themselves sideways to create stripes. However, Turing's original theory has many modern descendants, each of which attempts to address such deficiencies.

What about the *shapes* of animals? Shape and pattern are two aspects of the same thing—*morphology,* form in its most general sense. The changes of morphology that occur during biological development of an organism are called "morphogenesis." Turing *and* his critics generally agree that both shape and pattern seem to be set up in embryos by the same mechanism, a prepattern of chemical changes that waits for the appropriate stage of development and then triggers either pigments, to create pattern, or cellular changes, to create shape. The disagreement comes when we ask for the detailed mechanism that sets up the prepatterns. Not surprisingly, many of those changes have a genetic component—particular genes switch on simultaneously in row upon row of cells, stimulating the production of pigment, or causing the cells to modify their form and their mechanical or chemical properties. Nonetheless, the patterns of activity of the genes are taken, virtually unchanged, from Turing's mathematical pattern book.

So it looks as if DNA guides morphogenesis along the appropriate lines—but the response of the organism involves the laws of physics and chemistry, too. We know a lot about DNA, especially how it makes proteins, but we know far less about how those proteins and other DNA products are marshaled together to create a complete organism. It's frustrating. In fact, the problem of biological development is one of the biggest challenges facing science. How do organisms regulate and control their own growth patterns? What defines an animal's body plan? How is its form transferred from the DNA drawing board to the developmental assembly line? How does the living material of the growing organism know where—and what—to grow? The solutions for these questions involve chemistry, biology, physics, *and* mathematics—not just one of them.

Slowly, we are learning more and more about the processes that control biological development—both their genetics and their mathematics. In the 1980s, geneticists discovered that many different genes of the fruit fly *Drosophila* are preceded by a common DNA sequence, about 180 bases long. Moreover, the same sequence is found in many genes in numerous other organisms. Because this sequence has survived a lot of evolutionary change, it presumably does something so important that no changes to it can be tolerated by natural selection. The sequence is called a "homeobox." It looks as if the homeobox is some kind of molecular controller for genes, switching them on or off. It operates one level higher than the protein-making genes: Instead of controlling proteins, it controls other genes. It is possible to use fluorescent dyes to observe which genes in a chromosome are switched on, active. Some 1990s research shows that in appropriate circumstances, the active genes of the fruit fly larva will form striped regions (Figure 56). Later, the larva divides into segments along the edges of these stripes. This process clearly demonstrates a link between the organism's gene activity and its overall form and pattern. In 1986, Andrew Lumsden showed that the same types of patterns show up in the neural tubes of embryonic mice, so the experimental evidence is not confined to fruit flies. The genes that specify the structure of the embryo switch on in a spatially patterned manner, controlled by homeobox genes. Nonetheless, genes generally act in partnership with physics—or, in this case, chemistry—so homeoboxes are by no means the final answer.

Figure 56
Striped patterns of gene activity in a fruit fly larva.

Instead of looking at the genome to see how various pattern-making genes are switched on and off, Turing focused on the generalities of pattern formation. He realized that in a developing organism, a suitable chemical or chemicals, obeying his reaction-diffusion equations, could set up a prepattern that could subsequently be parlayed into markings or features of the animal's shape.[4] He called such a chemical a "morphogen," and his calculations led him to the discovery that a system of morphogens, reacting together and diffusing through tissue, can explain the formation of patterns. You don't need to put in the patterns to start with; they form spontaneously through the operation of the laws of physics and chemistry.

In fact, the patterns arise by symmetry breaking. There is always a trivial solution of the equations, a uniform state in which the chemical concentrations are the same anywhere. If this were a prepattern, it would lead to an organism that was the same all over. So the uniform state does not correspond to a significant pattern. However, it *is* significant for what it can give rise to. Because the uniform state is unstable, any slight lack of uniformity in the distribution of chemicals will grow rapidly. It might seem that this would produce just a random patchwork of chemicals, but the patches are coupled together by diffusion, and Turing discovered that this causes them to arrange themselves into coherent spatial patterns resembling spots, stripes, and other geometric textures (Figure 57).

Turing's theory had the right kind of feel, but it relied on invoking the existence of those special chemicals, the morphogens. When biologists looked for these chemicals, they couldn't find them. However, it's a very delicate task to find particular chemicals in an organism, especially when

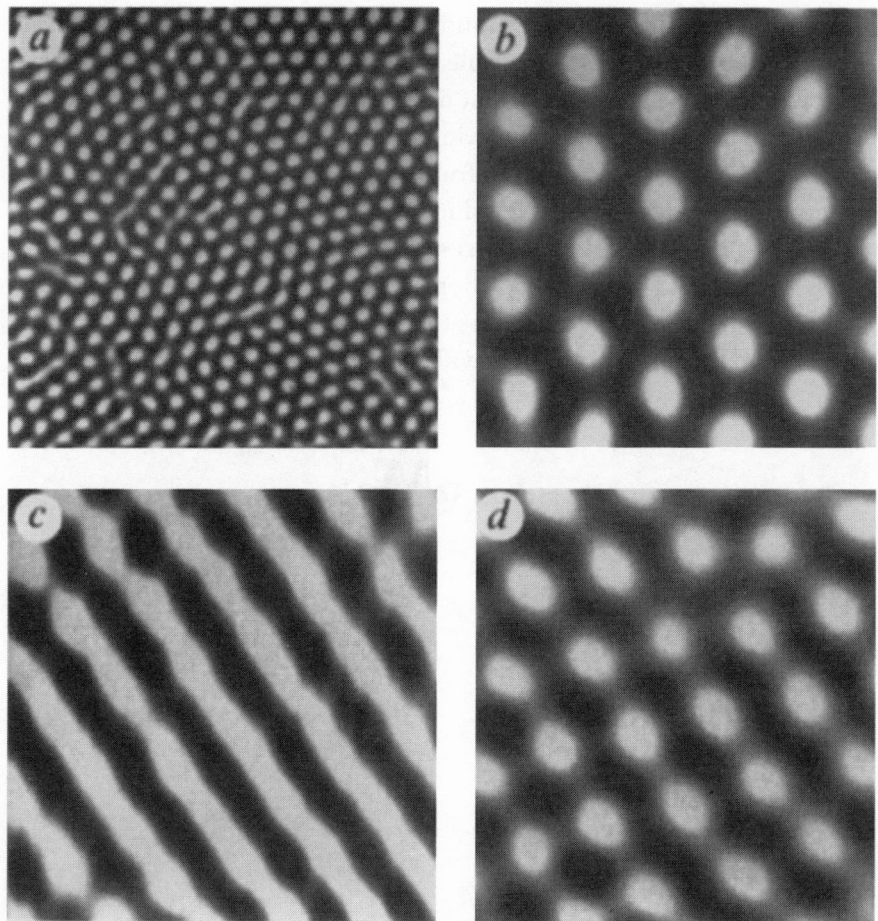

Figure 57
Experimental patterns created by reaction-diffusion, showing hexagons, spots, stripes, and spotty stripes.

you have no idea what the chemicals might be, so the biologists' lack of success wasn't conclusive. Chemists did manage to establish that Turing's mathematics applied to real chemicals. A good example of real-world Turing patterns is the BZ reaction, with its concentric rings and rotating spirals.

Concentric rings and rotating spirals aren't common on animals, but reaction-diffusion equations can also produce a huge range of other patterns—such as stripes, spots, dappling—that do appear on animals. Jim Murray[5] has applied equations similar to Turing's to the formation of dappling on giraffes (Figure 58) and of stripes on zebras and on big cats (Figure 59), proving the memorable theorem that "an animal can have a spotted body and a striped tail, but not a striped body and a spotted tail." Another apparent triumph for Turing patterns was the work of Maynard Smith, showing that hairs on the fruit fly occur in a variety of Turinglike patterns, the genetic variants of which are *also* Turing patterns. This patterning is hard to explain if patterns are arbitrary consequences of DNA codes—why should natural selection prefer Turing patterns?

However, these initial successes turned to failure, as other morphological systems, such as feather development, fell apart under scrutiny.

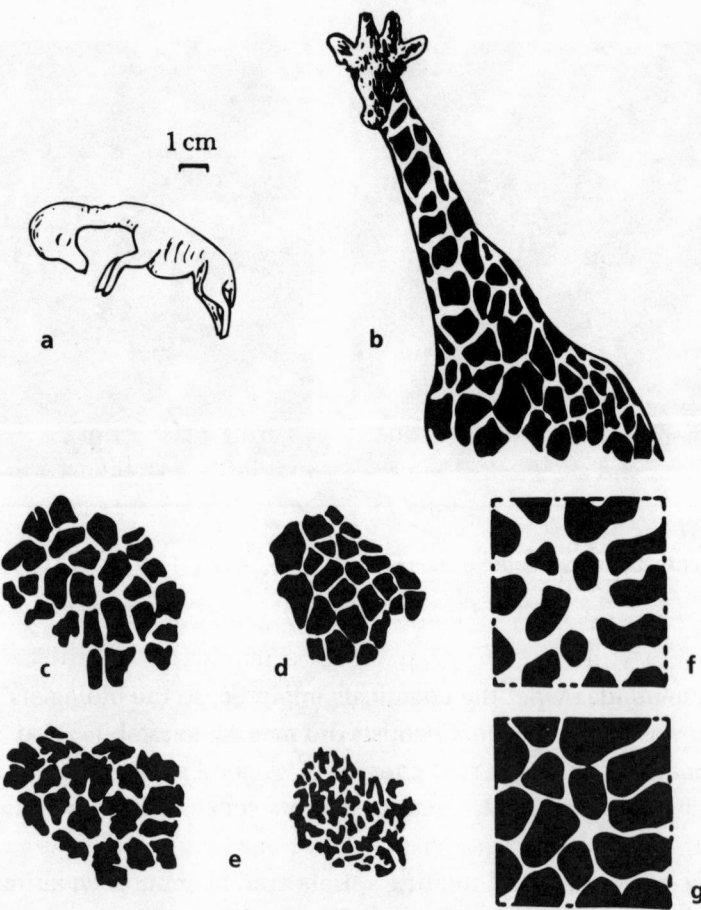

Figure 58
Simulations of dappling.

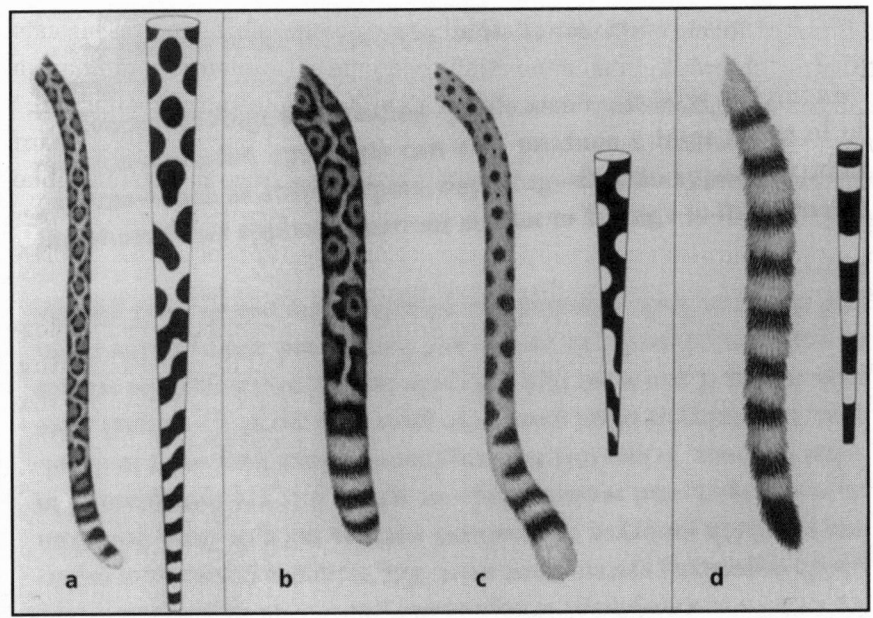

Figure 59
Simulations of spots and stripes on cats' tails: (a) leopard, (b) jaguar, (c) cheetah, and (d) genet.

When feathers are grown at different temperatures, the observed quantitative changes in their patterns do not fit Turing's equations. The discovery of DNA and advances in genetics posed further problems for Turing's theories. The fruit fly *Drosophila* is one of the geneticist's favorite experimental animals because it breeds rapidly, can be kept easily in the laboratory, and exhibits a huge range of different forms. An enormous amount is now known about fruit fly genetics, and it looks as if *Drosophila* builds striped patterns in a manner quite different from Turing's model. Indeed, evidence suggest that the fruit fly implements genetic instructions to build the stripes *one at a time*—quite unlike the reaction-diffusion mechanism, which lays down the entire prepattern in one go.[6]

By the 1970s, most biologists had become bored with finding Turing patterns that weren't, and they had moved on. Instead, they concentrated on the DNA code and its implementation, and they came up with an alternative theory, which I describe after a brief detour with butterflies, mermaid's caps, and angelfish. Mathematicians were less worried about the failures of Turing's equations because they saw those equations as just one example chosen from an almost infinite range of possible equations with similar properties. They realized that all systems *like* Turing's—including more elaborate mechanochemical equations describing the interaction of chemical changes and tissue growth—would produce the same general range of patterns. To them, the specifics of particular equations were unimportant; what mattered was the common features of the whole class of equations, which were the key to the general problem of pattern formation. So, as the twentieth century draws to a close, Turing's ideas are once more coming back into vogue, but in a more subtle guise. Mathematicians are introducing models that are closer to real biology, but that exploit the same symmetry-breaking mechanism of pattern formation. One application is to the markings of butterflies.

There are nearly a million different species of butterflies and moths, scattered throughout the world. They range from tiny, dull specks to soaring many-hued marvels. Nonetheless, a lot of order is hidden in their riotous displays, and in the 1920s, German naturalists B. N. Schwanwitsch and F. Suffert classified the patterns into a fairly small list of basic elements, such as spots, bands, borders, and so on. In 1981, Murray formulated mathematical equations for the patterning of butterflies, the predictions of which agreed with experiments on the developing insects.[7] His model combines Turing's reaction-diffusion systems with genetic switching. The morphogen is released from the edges of the butterfly's wing and diffuses toward the center; as it diffuses and decays, it switches on certain genes; the genes then release their own chemical products, which also react and diffuse. As these complex waves of chemical interaction spread across the surface of the wing, they lay down a prepattern that later determines the butterfly's markings.

Brian Goodwin has introduced another variant on Turing's scheme, a model of development in which, as a creature develops, its older parts become relatively fixed in form.[8] If the creature begins with circular symmetry, then the initial stages of growth also have circular symmetry; and that symmetry may become a stable pattern as the creature ages. Eventually, the symmetry breaks, and then the creature develops a number of equally spaced bulges that grow into branches or tentacles or petals. So the symmetry of the creature changes as you cast your eyes along the tissue from old to new. Goodwin's equations successfully describe one key step in the morphogenesis of *Acetabularia*, a single-celled marine alga, which follows a similar pattern. The creature begins as a spherical egg, which puts out a rootlike structure and a stalk. The stalk grows and produces a ring of small hairs, termed a *whorl*. The tip continues growing from the center of the whorl, producing more whorls; then it develops a capped structure, which gives it its common name, "mermaid's cap."

The new theories do the job fine—but in this area of science, old theories never die. Some of the early objections to Turing's original scheme are losing their relevance. For instance, it has long been assumed that Turing's equations have a tendency to produce moving patterns much more readily than stationary ones.[9] Certainly, this tendency is a feature of many Turinglike systems—both mathematical and experimental. Obviously, the patterns on living organisms *don't* move . . . big problem.

But . . . sometimes the patterns on organisms *do* move. They move rather slowly, which is why we don't generally notice movement, but they move. The organism that caused this latest change in thinking is one I mentioned earlier: a small tropical marine fish, the angelfish *Pomacanthus*. Juveniles are about 2 cm (less than 1 inch) long, adults three to four times that length. There are many different *Pomacanthus (P.)* species, and they exhibit a variety of patterns. *P. semicirculatus,* for example, has curved stripes that run vertically down the side of its body, whereas *P. imperator* has horizontal stripes along the length of its body. Over time, the pattern of stripes on each fish changes. This change is especially evident for *P. semicirculatus,* where the young fish have only 3 stripes, but the adults have 12 or more: Somehow or other, the number of stripes has to increase as the fish develops. In fact, the pattern changes in a rather curious manner. Start with a juvenile fish having 3 stripes, and watch it grow. At first, the stripes expand with the fish, becoming more widely spaced; this is what you would expect if the pattern were laid down once and for all. Then, suddenly, new stripes begin to appear between the original ones, restoring the original spacing. At first, the new stripes are thinner than the old ones, but they gradually thicken. When the body length reaches about 8 cm (a little over 3 inches), the process is repeated a second time.

Kondo and Asai have modeled this sequence of changes using reaction-diffusion equations.[10] Their model involves just two chemicals and assumes that the underlying tissue consists of a row of cells, some of which duplicate every so often. Their results, shown in computer simulation in Figure 60, reveal a natural pattern of stripes (the peaks in the figure), which widen, without changing their number, until the tissue

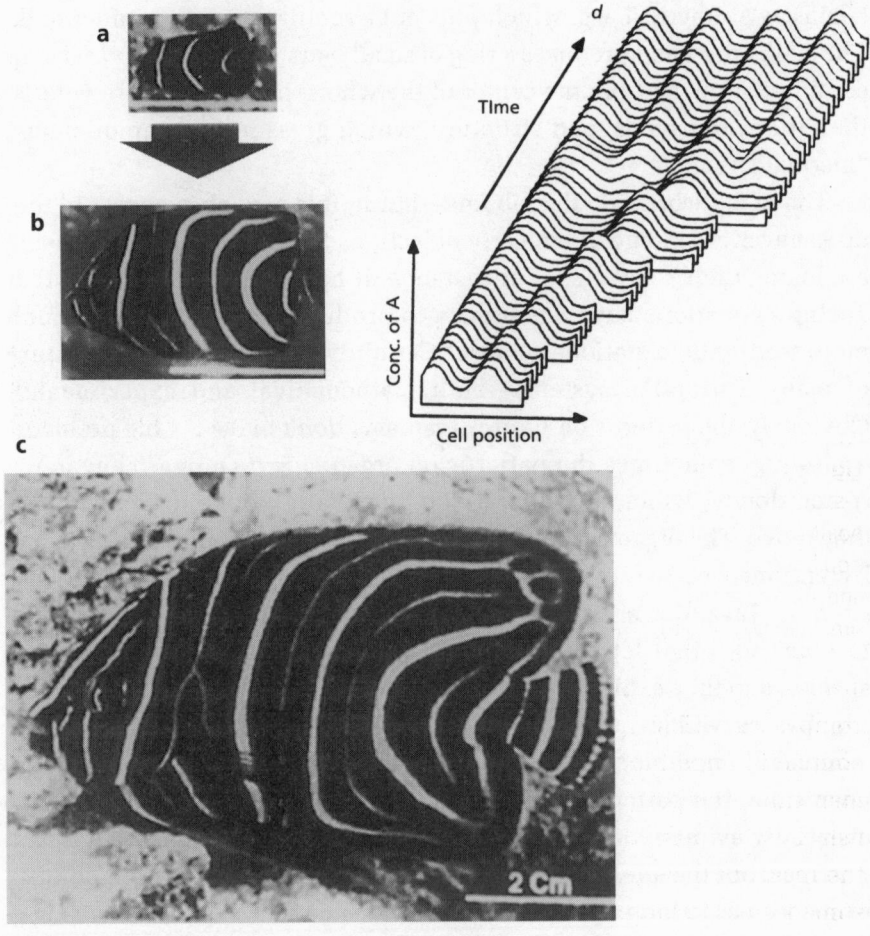

Figure 60
Changes in the number of stripes in a Turing pattern.

becomes sufficiently large, at which point the number of waves doubles, with new stripes appearing between the old ones.

An even more dramatic scenario arises in the horizontal striping of *P. imperator*. It also develops additional stripes as it grows, but various of the stripes appear to unzip and split into two. This type of wave rearrangement is known to physicists by the term dislocation; dislocation is widely observed in a variety of systems, including reaction-diffusion systems. To say that the stripe unzips is a slight simplification because it suggests that a single stripe turns into two by developing a Y-shaped branch-point. It *can* happen this way (Figure 61), but there are also more

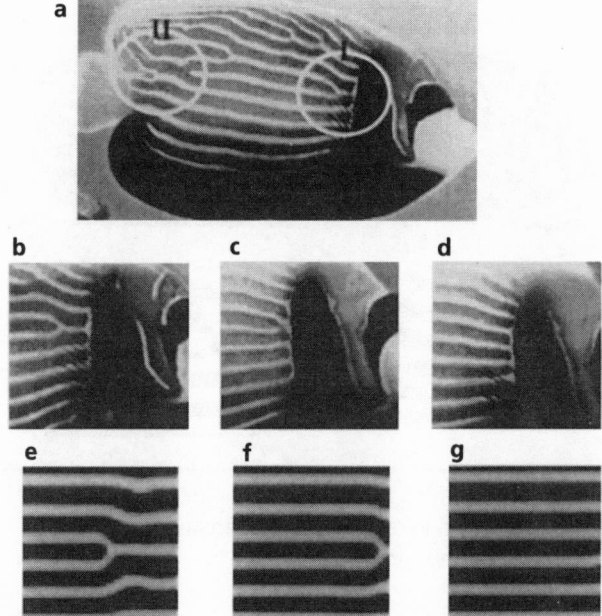

Figure 61
Y-shaped branch
point in an angelfish
(top and middle)
and Turing-pattern
simulations (bottom).

complicated dislocations in which stripes rearrange themselves by disconnecting and reconnecting, and Kondo and Asai saw these arrangements, too (Figure 62). They used the observed spacings of stripes in the fish to estimate the diffusion rates for their hypothetical morphogens, and the results are within the range you'd expect if each morphogen were some kind of protein molecule. The evidence may not be watertight—but if what's going on is *not* governed by a reaction-diffusion equation, it must be governed by something very similar.

Now let's go back a few pages, to where the biologists were abandoning Turing's idea, and pick up the second thread to the tale. We do not yet know how development is organized, but it is clear that some very adaptable organizing system must exist. The evidence comes from experiments in which embryos of creatures such as mice or frogs are taken apart, rearranged, and reassembled—or where bits are transplanted to see what happens. When this rearrangement is done very early in development, the embryos often manage to grow into perfectly normal mice or frogs. It is as if you went into the cycle factory, exchanged the wheels of the embryonic bike for its saddle and handlebars, wrapped the chain around the front fork, attached the wheels to the pedals, and put the brakes where the gears usually go—and then watched in incredulity as

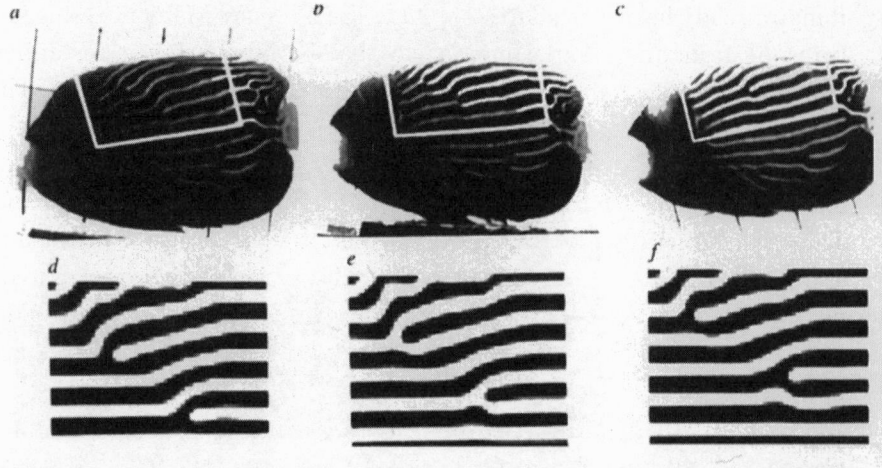

Figure 62
Disconnections and reconnections in an angelfish (top)
and Turing-pattern simulations (bottom).

the bicycle calmly rearranged its parts back where they ought to have been in the first place. On the other hand, transplant experiments carried out later in development usually give rise to substantial differences in the resulting organism—extra bones, deformities, feathers in the wrong place. . . .

The most successful theory of the development of the embryo is Lewis Wolpert's theory of "positional information."[11] This theory maintains that the cells in the embryo seem to know where they are at all stages of development because of chemical signals. They can then use this information to follow different paths, depending on where they are. Cells that are in the region where a heart ought to be, for instance, can consult their DNA for the instructions for building a heart, and so on. If they don't know where they are, they can't make the right decision. In short, cells possess both a *map* (positional information) and a *book* (their DNA code). Depending on where they lie on the map, they consult the book, which tells them what to do at that location.

The theory of positional information is an elegant and appealing theory, and in many respects, it must be close to the truth. It has a particularly good track record of explaining the results of transplant experiments. Tissue moved to a new position retains its book, but its map gets confused. The transplanted tissue may respond to the new map, or it may already have started to respond to the old map. Many puzzling features of transplant experiments can be explained in this manner. However, the mechanism of positional information is rather rigid: The map is predetermined, and the construction steps are all recorded in the book before work starts. More contemporary alternative theories, which also fit the results of transplant experiments, take a more dynamic view: The organism still has a book, but it builds its own map as it grows, and a lot of the time, it does not need to consult the book to know what to do next. While such a process may appear more mysterious, it is closer to the natural flow of developmental dynamics, and there are reasons to believe that it may come closer to capturing the realities of development.

Wolpert's theory, like Turing's, is based on the concept of a morphogen, a kind of chemical signal—but it employs this signal in a very different way. In the early part of this century, geneticist Thomas Hunt Morgan suggested that organisms such as worms, which can regenerate themselves if they are chopped into pieces, do so by means of chemical signals. He suggested that the cell senses the *gradient* of chemical concentration—the direction along which the amount of that particular chemical increases—and that gradient, in effect, tells the cell which end is which, and where it is in relation to the front and back ends of the worm. Wolpert's idea is that gradients in the concentration of a whole system of morphogens can tell a cell *exactly* where it is in the organism. With this information, all the cell has to do is consult its genetic instruction book and do whatever the book tells it. So versatile is this system that it can produce any pattern whatsoever—circles, squares, or a copy of the Mona Lisa.

Analogously, a computer screen can create any image you want if it is supplied with a pixel-by-pixel list of the required colors and intensities. *Any* pattern can be approximated by colored cells on a square grid. A system of three distinct chemical gradients can divide the space occupied by an organism into three-dimensional pixels, so any desired pattern or form can be specified in the DNA codebook. Of course, the real method doesn't have to be quite *that* neat and tidy: The organism might employ more than three chemicals, for added checks and to avoid mistakes. Also, there is no special reason to use a rectilinear Cartesian grid. Combine this system with a sufficiently detailed book of instructions, and you don't just get the possibility of marking the Mona Lisa on an animal—you could create the Venus de Milo in three dimensions, missing arms and all.

In Turing's theory, the patterns are created by the laws of physics and chemistry. A chemical does not have to know where it is to know what to do; it just does it. In Wolpert's theory, in contrast, the patterns are created by what is written in the book; the only role played by the morphogen is to tell the cell where it is on its map. Figure 63 shows how a gradient-dependent scheme could set up a striped pattern of tissue. For simplicity, we show only one dimension of space, broadened into a thin strip of tissue to make the pattern clear. The reality is three-dimensional. The idea is that there should be many critical levels of the concentration. The cell's DNA book knows that if the concentration is between levels 0 and 1, then it should not activate genes producing black pigment; if it is between levels 1 and 2, then it *should* activate genes producing black pigment; and so on—activate between an odd level and the next one up, don't activate between an even level and the next one up. It's a simple, tidy scheme. All of the complexity, versatility, and subtlety goes into the book of instructions, and you don't need any fancy mathematics. Why should organisms follow the Turing approach when there's a more direct and more versatile one available?

It's a good argument, but the spontaneous patterns in Turing's approach come without any prompting or expenditure of energy, thanks to the laws of physics and chemistry. Why should nature voluntarily *decline* to use what's so readily available? The truth, no doubt, is that organisms employ a combination of both DNA and natural laws. Let's see what's actually known about the question.

There is plenty of circumstantial evidence in favor of morphogens, but only in the 1970s did plausible chemicals make their appearance on the scientific stage. In vertebrates, the most plausible suspect is a molecule

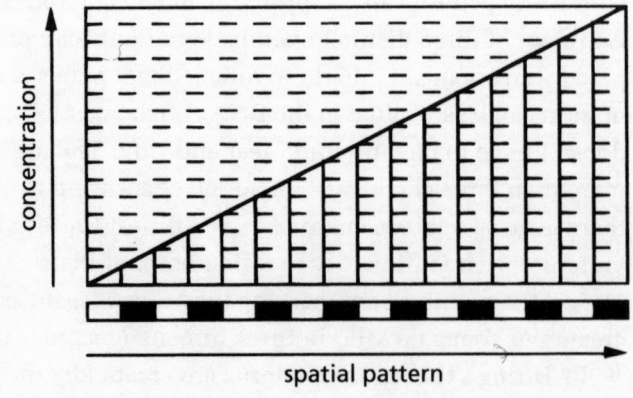

Figure 63
How to generate stripes from a single gradient.

known as retinoic acid—but whether it really *is* a morphogen is still disputed. The evidence *for* its morphogenicity includes some remarkable effects observed in the growing wing of a chick embryo. The normal gradient of retinoic acid is high at the rear end, low at the front, and decreases more or less steadily in between. The corresponding pattern of bones in the wing is shown in Figure 64 (top). Digits 4, 3, and 2 develop at decreasing levels of concentration of retinoic acid. If part of the growing limb is transplanted, to create a retinoic acid distribution that is high at both ends and low in the middle, then an abnormal pattern of digits results (Figure 64, bottom). The digits double up, and one set—the one in which the gradient of retinoic acid is reversed—is a mirror image of the other set: 4 3 2 2 3 4.

Case closed? Not at all—biology is infinitely subtle. More recent experiments suggest that the second source of retinoic acid may act as a trigger, rather than a gradient. It may just switch on whatever other system in the embryo is responsible for limb development. Moreover, some genes that—according to Wolpert's theory—*ought* to respond to retinoic acid, do not always do so. All this experimentation makes the theory excellent science, frontier stuff: The kind of interpretational problems that are occurring are what you expect at the frontier, and they will provide a route to eventual progress and understanding. Nonetheless, for now, they

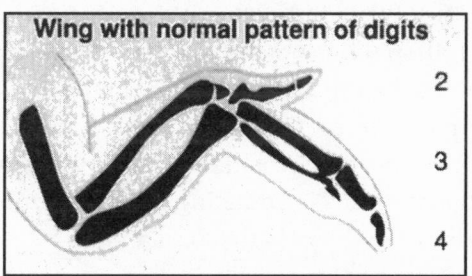

Figure 64
Development of the chicken wing in a normal gradient of retinoic acid (top) and in an abnormal two-way gradient (bottom).

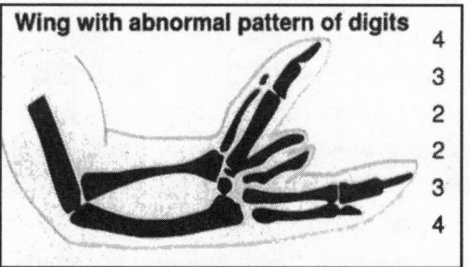

muddy the scientific waters and prevent us from proclaiming the discovery of a specific morphogen.

Independent of biology, there are several other objections to Wolpert's scheme. The first one, ironically, is that it is *too* versatile. While some animal patterns and forms are pretty much arbitrary—markings on birds, for instance—most of them seem to be drawn from a relatively small catalog of structures—spots, stripes, patches, blisters, spheres, discs, tubes, fronds, whatever. These basic forms have a distinctly mathematical appearance: maybe not regular, but constrained. If nature can make Mona Lisas as readily as stripes, why do so many animals have stripes?

The next worry is that while some kind of chemical-coordinate grid and a pixel-by-pixel coloring book is the kind of thing that might be designed by a human engineer, it is a bit too tidy for biology. A computer screen can *represent* a fruit fly larva pixel-by-pixel—but that doesn't mean that nature grows fruit flies *using* such a description.

A third objection is one of precision. Wolpert often illustrates his theory with a diagram rather like Figure 63, but having only two critical levels, chopping the morphogen concentration into three adjacent ranges. With these ranges corresponding to red, white, and blue, the result is a French flag. I'm happy to buy French flags, with three big stripes—but how about a zebra? You have to pack an awful lot of stripes into one gradient. The slightest error in distributing the morphogen—or in reading its concentration—would put the stripes all over the place, some thick, some thin.[12] If you tried to evolve a Mona Lisa using gradients, I'm pretty sure you'd end up with something more like a Jackson Pollock. So, to be blunt, I don't buy gradients except as a simple method that might be used for very simple patterns. Of course, those patterns might be simple parts of something more complicated, but then you have to organize the parts as well, and that means introducing even more morphogens.

So far, all of the theories discussed have a common defect: the idea that the position of any specific part of an organism can be specified uniquely, once and for all. Reaction-diffusion equations are defined in fixed regions of space; Wolpert's scheme employs a predetermined grid. However, organisms change shape—drastically—as they grow. To some extent, the defect can be repaired by making the map or grid change shape as the organism grows—but keeping track of position through all these changes is decidedly nontrivial—especially because bits may die or otherwise be discarded during growth. What would seem to fit the bill better is an approach in which the shapes and positions are not built in to begin with, but emerge from the growth process itself, through spontaneous self-organization.

How can we achieve this? Here's one suggestion, a simplified version of a more sophisticated system introduced by Pankaj Agarwal, under the name cell programming language (CPL).[13] I don't intend it to be taken too seriously as a contender for a viable theory, but it does demonstrate that there are systems that are far less sensitive to errors than gradients are. The starting point is that you don't need to divide an animal into pixels— it is already built that way, only with pic-*cells*. Cells communicate with their neighbors by chemical means, transmitting or receiving molecules across their adjacent membranes. This structure is ideal for implementing rule-based systems of instructions.

Here's an example. For simplicity, imagine we have a line of gray cells that can be switched to two other states—black and white, say. (I introduce gray only to distinguish cells not yet switched on by the chemical signaling system.) Equip them with just one rule: If the cell to your left is white, become black, and vice versa. Start by switching the leftmost cell to black, and watch what happens (Figure 65, top). Obvious, isn't it? Not only that: The system is *extremely* robust. The individual cells could change size, shape, whatever. As long as their relative positions remain the same, you'll get the same pattern, so small errors in building the organism have no serious effect on the pattern. In other words, discrete switching rules produce more reliable results than does a continuous gradient. But there's more: Suppose the instruction affects not color but size and shape—with "black" replaced by "stay the same size" and "white" by "shrink." Then you get structured tissue, like Figure 65, bottom. Again, the pattern that the rules produce is not sensitive to small errors.

It takes very little imagination to see the possibilities for this kind of organism-building system. It could easily be implemented by real cells, with adjacent cells signaling chemically to each other across their common membranes. Moreover, instead of requiring an entire catalog of gradient thresholds, this method of growing an organism's pattern and form employs just a few simple rules. It *looks* organic. This observation leads to my final worry about the gradient method: It is inefficient. It takes huge amounts of information to determine not just Mona Lisas, but also any pattern with the same number of pixels as a Mona Lisa. There are more cells in the human body than there are bases in the human genome, by several orders of magnitude; so the book can't deal with every place on the map. The best way to reduce information requirements drastically is

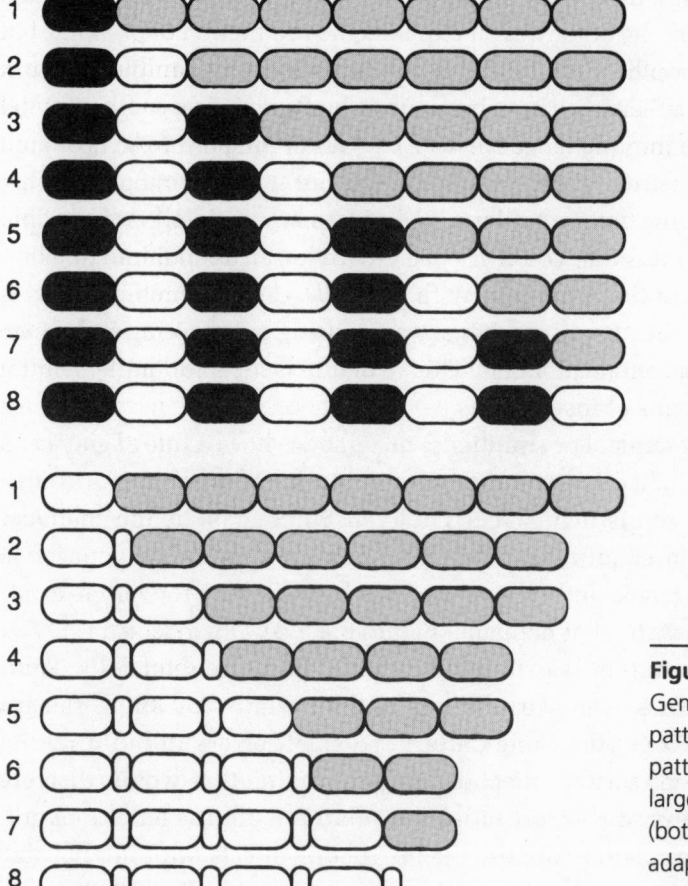

Figure 65

Generating a striped pattern (top) and a pattern of alternating large and small cells (bottom) in an adaptable cellular automaton.

not to find more clever ways to encode lists of data: It is to replace those lists by simple rules that can generate the data.

This suggestion leads to a prediction: The patterns and forms most commonly observed in nature will be the ones that can be generated by the simplest cellular rule schemes. Uniform regions are perhaps the *simplest* of all, then stripey forms, then spotty ones . . . after that I can't guess sensibly. Not knowing the language of the rules, there's no very good way to decide. Nonetheless, this system has the *virtue* that it does *not* make all patterns possible—or at least not equally likely. Also, the rules make use of the natural patterns inherent in cellular systems, exploiting what is already available: That's where their compact descriptions come from.

Early mathematical equations for development were too far removed from real biology to provide accurate models. The current emphasis on DNA goes too far the other way: It explains the production of proteins, but it does not adequately explain how the proteins are assembled to form an organism, or—crucially—why nature so often prefers mathematical patterns. To see the difference between the two approaches, and how both fall short of reality, imagine a vehicle (corresponding to a developing organism) driving through a landscape (respresenting all the possible forms that the organism might take, with valleys corresponding to common forms and peaks to highly unlikely ones). In models such as Turing's, once you have set the vehicle rolling, it has to follow the contours of the landscape. It can't suddenly decide to change direction and head uphill if the natural dynamic is to continue straight ahead into the nearest valley. In contrast, the current view of the role of DNA sees development as an arbitrary series of instructions: "turn left, then straight ahead 100 meters, then turn right; stop for 10 seconds; reverse 5 meters; turn left. . . . " *Any* destination is possible, given the right instructions, and no particular destination is preferred.

The true picture, however, must *combine* genetic switching instructions and free-running dynamics. If a car driving through a fixed landscape follows an arbitrary series of instructions, it is very likely to drive into a lake or off the edge of a cliff, and it has little chance of reaching the top of a mountain. On the other hand, a car with a driver has more freedom in selecting destinations than does a free-running vehicle without any controls. In the same manner, an organism cannot take up any form at all: Its morphology is constrained by its dynamics—the laws of physics and chemistry—as well as by its DNA instructions. Still, the DNA instructions can make arbitrary choices among several different lines of development that are consistent with the dynamic laws. The new mathematical models are finally beginning to put these two aspects of development together. DNA alone does not control development—nor do dynamics alone. Development requires both, interacting with each other, like a landscape that changes shape according to the traffic that passes through it.

8

The Peacock's Tale

The jewelled splendour of the peacock and the
humming-bird, and the less effulgent glory of the
lyre-bird and the Argus pheasant, are ascribed to the
unquestioned prevalence of vanity in the one sex and
wantonness in the other.

D'Arcy Thompson, *On Growth and Form,* Chapter XVI

For pride, little can outdo a strutting peacock—and
he has a great deal to be proud of, has he not? His neatly crowned head,
his beautiful iridescent blue body . . . and, above all, that magnificent
tail—a tail that can spread out like a gigantic fan, green, gaudy, bejew-
eled with striking round spots. Impressive.

The peahens, too, are impressed. And yet, the peacock is a silly bird.
Why waste all that effort dragging such a ridiculously overblown tail
around? It doesn't help you to fly. In fact, it makes flying rather difficult.
Nonetheless, the poor peacock *has* to drag his beautiful, clumsy tail
around. Why? Because the peahens are impressed.

If you are a peacock, you have to impress the females in order to mate and thereby pass your genes on to future generations. The idiotic splendor of the peacock's tail, as Darwin himself recognized, is just one example of a general process: sexual selection. The usual way to explain this is that if, by accident, a preponderance of females happen to possess a gene that leads them to prefer to mate with males with a particular feature, then males lacking that feature will have less success at the mating game. In short order, that tiny imbalance of female genes drives the male populations to wild excesses of plumage or behavior. To complete the feedback loop, females who happen not to like that particular embellishment in their males find it harder and harder to locate a mate that lacks it. So the females, too, become ever more enamored of whatever accidental feature happened to be introduced when the process began.

Research, however, indicates that the female's preferences do not always depend on chance. It is a strange story, a tale of senses, symmetry, and sex. Senses come into the picture because it is the female's *perception* of the male that leads her to choose her mate. Symmetry is involved because senses work effectively only if they pick up genuine patterns from the outside world, and the outside world is based on physics, the laws of which are deeply symmetric. The role of sex is obvious.

It is also a story of *emergence,* the term philosophers use for phenomena in which "the whole is greater than the sum of its parts." More precisely, emergent phenomena occur when a system composed of many individual elements exhibits collective behavior that seems not to be built into the individuals in any obvious or explicit manner. For instance, the human brain comprises innumerable nerve cells, yet a nerve cell seems not to possess the tinest smidgeon of intelligence. Our intelligence *emerges* from the interaction of those billions of nerve cells—you won't find it if you look at them one at a time.

Most of the senses in complex animals are emergent. And so—sometimes—are the sexual preferences of females. What laws govern the form and function of senses? Are those laws genetic, mathematical, or a combination of both?

Sensory systems generally have a very mathematical structure. Why? Senses have to process incoming data rapidly and reliably to extract information about the outside world. To do this, they must be engineered rather carefully. They must respect the underlying patterns of the outside world. For example, because external objects can move around, a visual sense will function much more effectively if it can recognize the same object at different distances and in different orientations. This requirement has important implications for how the brain processes visual signals. Sensory organs wouldn't work very well *unless* there were a considerable element of mathematical regularity in their design. On the other hand, because these organs are the products of evolution, we shouldn't expect their mathematical patterns to be perfect, or without exception.

We all know that the common element, in brains and sensory perception, is the nerve cell. We also know that brains are lots of nerve cells connected together in a huge network—there are other ingredients, too, but the nerve cells are of paramount importance (Figure 66). Nerves also connect sense organs to the brain, conveying sensory inputs, which the brain processes into perceptions. It is less well known that nerve cells also run the other way, from the brain to the sense organs, so that the brain can fine-tune its senses to pick up whatever it considers to be most important. Perception is a two-way street.

Nerve cells form long, thin fibers that carry electrical signals. These fibers are a bit like telephone wires, but although this image is convenient, the biological transmission of electrical signals differs from the nonbiological transmission of signals—as Alan Hodgkin and Andrew Huxley discovered in 1947 by studying a squid nerve cell. Their work, which

Figure 66
Networks of nerve cells in the brain.

won the Nobel prize, was an early triumph of biomathematics; it depended on writing down equations for the transmission of the nerve impulse, known to this day as the Hodgkin–Huxley equations. They found that a nerve cell transmits a signal in the same way that a spark travels along a fuse. The signal involves not free electrons, but ions— atoms with an extra electron, or a missing one, which gives them a positive or a negative electrical charge. The ions do not travel *along* the nerve, but *across* its cell membrane. Electrical activity at one position along the nerve triggers activity in the next section, just as burning in one segment of a fuse triggers burning in an adjacent segment. The only thing that travels is the location of the activity.

A single nerve cell can send signals and can respond to the signals it receives, but (as far as we know) it can't do much *with* them. Effective sensory systems, and especially brains, need to be able to carry out processes that manipulate electrical signals. In a sense, they must be able to carry out computations—though not necessarily in the same sense that a computer carries out computations. Sensory systems acquire computational abilities by hooking up into networks, forming complex biological equivalents of electronic circuits. How hard is it to accomplish computation? Does evolution have a lot of work to do? In 1982, John Hopfield invented a simple mathematical model of a network of nerve cells, called a "neural net," and he showed that if you just hooked a lot of simple neural units together, they acquired computational abilities.[1] You could design networks for particular purposes, just as an engineer designs electronic circuits for hi-fis and computers. More surprisingly, though, you could also just throw together a random assembly of neurons; despite their lack of design, such nets could *still* perform meaningful computations—but you wouldn't find out *which* computation until you built the network.

Hopfield called this phenomenon "emergent computation." It means that nets with raw computational ability can arise spontaneously through the workings of ordinary physics. Evolution will then select whichever nets can carry out computations that enhance the organism's survival ability, leading to specific computations of an increasingly sophisticated kind. Somewhere down that road lies the brain. So Hopfield's view was that the basics of a brain arise spontaneously, as a result of the general mathematics of neural nets.

Hopfield's model neurons are units that can be either active (on) or quiescent (off). The model neurons respond to incoming signals by switching between these states, according to the strength of the incoming signal. The network can learn from experience by adjusting the strengths of connections, according to a rule devised by physiologist Donald Hebb in 1949: "Cells that fire together, grow together." That is, connections among cells that are active at the same time tend to strengthen; connections among cells that are not often active together tend to weaken. Connections can be of two kinds: excitatory or inhibitory. A neuron will become active if it receives enough excitatory signals to exceed some chosen threshold, but no matter what the incoming signals are, a single inhibitory input will make it shut up shop. If you connect lots of neurons together, they can carry out more complex tasks. When one neuron becomes active—switches on—it sends signals to other neurons. The receiving neurons, in their turn, switch on, and a pattern of activity and inactivity cascades through the network. Computers work in just this manner, using microscopic electronic switches called "logic gates." A neural net can, in principle, do anything that a computer can.

A good way to explain the behavior of a Hopfield network is to represent its state geometrically. Imagine a space with one dimension for each neuron of the network—the network's phase space. A single point in the phase space represents the activity levels of all the neurons in the net; each coordinate of that point gives the level for the corresponding neuron. Suppose, for instance, that the point in phase space has coordinates (1,0,1,1, . . .). Then this says that neuron 1 is on, neuron 2 off, neuron 3 on, neuron 4 on, and so on. The strengths of the connections thus determine a mathematical landscape that sits over the top of phase space. The height of features in the landscape tells us how much computation the net has to do to work its way to the right answer. Down in the valleys, it gets to the answer very quickly; up on the peaks, the calculations go on for a long, long time. So the aim of the game is to get off the peaks and down into the valleys. The rules with which Hopfield equipped his networks achieve this by making the net behave rather like a ball rolling around in the landscape: The action moves away from peaks and settles into valleys. Just as the rolling ball cannot avoid homing in on a valley, so the net cannot avoid homing in on a solution to its computational problem—eventually.

Later, Hopfield and David Tank[2] extended the idea to allow neurons to have a continuum of states, rather than just on/off, and those networks turned out to have all of the computational power of the discrete on/off networks—but, as a bonus, they could solve more complicated problems more rapidly.

To summarize: There is a natural evolutionary route from universal mathematical patterns in the laws of physics to organs as complex as the brain—with sophisticated powers to process incoming signals, store them in memory, retrieve them, associate different memories with each other, control locomotion. . . . The raw ingredients that allow evolution to get started are spontaneous, emergent features of large networks—be they biological or mathematical. The computational abilities that Hopfield observed in large random networks were not built into individual neurons. The individual neurons were just switches. No, the computational abilities emerged from the manner in which the neurons were connected. The proof of emergence was that it was almost impossible to predict how any particular neural net would behave. All you could do was run it and find out—which meant that no straightforward procedure existed that could deduce the collective behavior from a circuit diagram—and that's another way to say "emergence." So universal mathematical features of neural nets provided a bag of computational devices for evolution to build on. And once evolution gains a toehold, it soon takes over completely; it's getting started that counts.

Of course this doesn't tell us exactly *how* the brain evolved, which is a much harder—and very interesting—problem. Neither does it give us any clues about how the brain's networks actually operate. The brain contains about 10 billion neurons, and each is connected, on average, to about 10,000 others. It is impossible to trace out the complete circuit diagram, and in practice, it is very hard to trace even a tiny part of it. Investigators therefore look to simpler organisms. Chris Sahley and Alan Gelperin[3] chose to study the common garden slug, *Limax maximus*. The slug's favorite food is carrots, but it dislikes quinine. Sahley and Gelperin discovered that they could make a slug dislike carrots, too, by ensuring that the smell of carrot was always accompanied by the taste of quinine. Moreover, once the slug had learned not to eat carrot because it was paired with quinine, then it would also avoid any food that was paired with carrot. So in some sense, the slug learned to be put off not just by quinine, but also by things that its nervous system *associated* with quinine. Sahley and Gelperin also worked out the general nature of the neural net that the slug uses to carry out these computations, and they simulated it with a Hopfield-style net (Figure 67). The model net faithfully captured the kind of learning that the slug displayed. So, although Hopfield nets are grossly simplified models of real neural computation, they can still behave like the real thing in important respects.

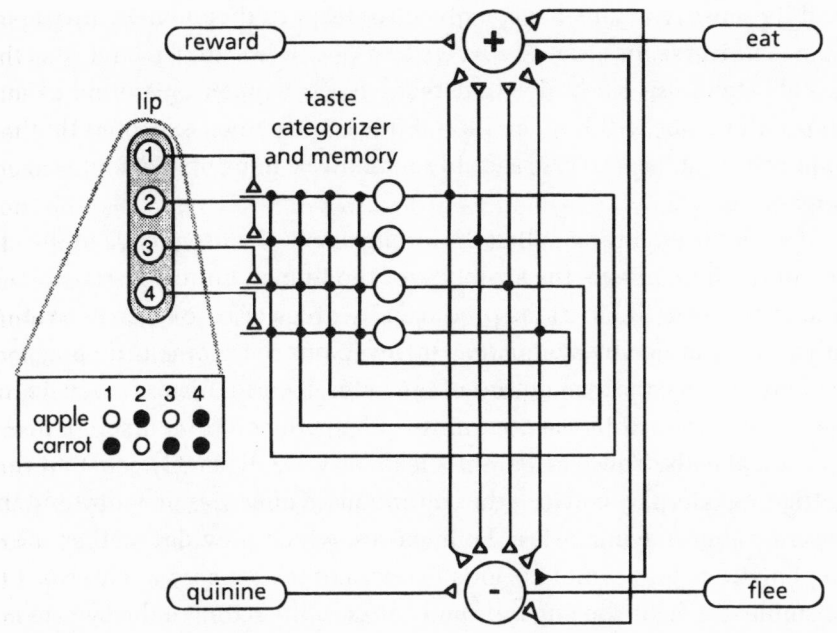

Figure 67
Model neural network for learning in the slug.

More sophisticated organisms have more sophisticated senses—hearing, vision, smell, taste, touch, a sense of temperature . . . even a sense of direction. The neural nets for such senses may have started out as small random circuits, but as they evolved, the changes that took place must have been affected very strongly by the constraints of the external universe. For example, if a prey organism's visual system can recognize a predator only when it is close up, that system doesn't greatly improve the prey's chances of survival. So vision must somehow recognize that a small image of a distant predator represents exactly the same external object as a large image of a nearby one.

Senses produce internal patterns of neural activity in brains; in order to be useful, those patterns must correspond, in some manner, to significant patterns in the outside world. Therefore, the neural nets used in sensory perception must be organized in a way that reflects the deep patterns of the external universe. Now, at its deepest level, the universe is based on symmetries. As we've seen already, symmetry governs the four forces of nature; quantum mechanics; the nature of space, time, matter, and radiation; and the form, origin, and destiny of our universe. The human brain and its sensory organs have evolved together in this wonderfully symmetric universe. In order to survive in a hostile world, we have evolved the ability to use our brains to detect what is happening around us and to predict what will happen next. So it is only to be expected that the symmetries of the universe should somehow be imprinted on our sensory apparatus.

I want to convince you that the arrangement of sensory cells in the retina of the human eye, the structure of the human brain's visual cortex, and even the hallucinations produced by drugs or by oxygen starvation are all determined by symmetry. However, our sense organs do not slavishly reproduce the symmetries of the natural world. Instead, they do the best they can, within the contraints of being made from discrete units, biological cells. These constraints lead, as we will shortly see, to a very intriguing interplay between the continuous symmetries of nature and the discrete almost-symmetries of our senses, which provides mathematical explanations for several curious features of the sensory apparatus. For example, the light-sensing rods and cones in the retina of the eye are not all identical. Instead, rods and cones near the center of the retina are very tiny and packed very closely together, while those farther out are larger and more sparsely distributed. This might be an accident of the development of the eye, but on closer inspection, it seems to happen for a reason. It makes it far easier for the visual system to recognize that an object seen at different distances is the *same* object. The common link between the external world and the structure of the eye is symmetry—here the dilational symmetry that magnifies or shrinks objects in the visual field as they move closer or farther away.

Let's quickly remind ourselves about the mathematics of symmetry. An object is symmetric if it remains unchanged after a *transformation*. There are several kinds of transformation: reflections, rotations, translations, and dilations (changes of scale). Dilations are "continuous" symmetries, meaning that you can scale an object by *any* amount. The same goes for rotations and translations, but not for reflections. However, our visual sense organs are made from discrete cells and cannot have continuous symmetries—under any symmetry transformation, those discrete units must click on or off through a definite distance or a definite angle, not through *any* distance or angle. Nature is therefore forced to approximate. Our visual system is a compromise, which replaces symmetries by *almost symmetries*—by this, I mean that transformations do not maintain the precise positions of all of the sensory cells, but the transformations move to positions that are, overall, always close to the previous position of some cell—probably a different one, but a cell nevertheless.

This concept of almost symmetry applies to the retina, where small cells are densely packed near the center of the visual field, and the cells farther out are larger. This structure is almost symmetric under rotations and dilations; in fact, it is as close to perfect symmetry as you can get with discrete units. This structure makes it easier for our brains to process incoming information from the eyes and to recognize that an object is the same whatever its orientation, or that an apparently large nearby object is really the same thing as an apparently smaller, more distant one. The translational symmetry of the external world is dealt with in a different manner: The brain points the eye toward the object, to bring it near the center of the visual field where it can be seen most clearly. When our eyes follow a moving object, we translate our field of vision to match the translation of the object. For this reason, the arrangement of cells in the retina has no need for translational symmetries. This is extremely fortunate, because no arrangement of discrete cells can have a good approximation to all three types of symmetry (not including reflections) at once. Mid-1990s research by computer scientists Simon Clippingdale and Roland Wilson and mathematician Peter Mason[4] has shown that an artificial neural network can be trained to learn how to form almost symmetric configurations by randomly exposing it to the corresponding symmetries of the external world and then modifying its form according to its response. Moreover, there is evidence that the positioning of cells in the retina is tuned during the development of an individual—perhaps by a similar training process.

The visual cortex—the part of the brain that receives and processes signals from the eyes—also has symmetries, but these are quite unlike those of the retina. In the visual cortex, translations dominate. As a result, our senses have to map the retina onto the cortex, and this is done by way of a mathematical transformation known as the "complex logarithm" (Figure 68). This transformation turns circles and spirals on the retina into straight lines across the cortex in various directions. Spirals are a very common type of hallucination, and if the patterns seen in hallucinations are transformed by this logarithmic mapping, they turn into systems of parallel lines, a fact first noted by Jack Cowan[5] (Figure 69). This observation tells us at once what must be the source of the hallucinations: Parallel waves of electrical activity are crossing the cortex like ocean waves rolling up a beach, and our visual senses are misinterpreting these

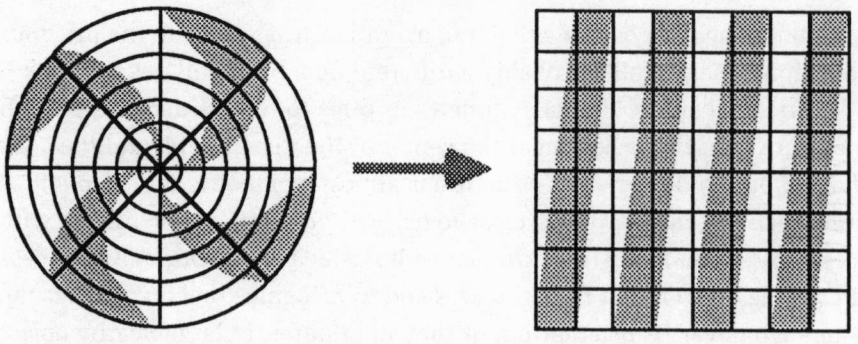

Figure 68
The complex logarithm transforms retinal images into patterns of activity in the visual cortex.

waves as if they were spiral signals hitting the retina. So our brain makes plane waves, but we think we see spirals. Reality is affecting our brains directly, instead of through our senses, and what we see differs from the reality that causes us to see it.

Symmetry, then, plays a key role in sensory perception because the neural networks that process sensory information have to be able to pick up the significant patterns of the external world—such as the geometry of space. It is no surprise, then, that symmetry shows up when it comes to females' sexual preferences. We saw earlier that Darwin realized that sexual reproduction opens up a new evolutionary phenomenon: sexual selection. Chance female preferences can become firmly established in the male—and the female—population, as they did for peacocks' tails. Darwin did not explain why such preferences might arise to begin with, but in 1930, Fisher pointed out that it could be the result of a runaway feedback loop. The ingredient that makes the process possible is a link between the females' perceptions and their choices of mates. Male genes for fancy tails and female genes for preferring fancy tails perpetually reinforce each other, and the process stops only when it is brought to a halt by something with an even stronger effect on reproductive success. The feedback is triggered by that accidental slight initial preference, and the whole story could have gone differently—with a preference for small tails, or curly beaks, or big feet instead. Birds of paradise—with their exotic, elaborate, polychromatic, and otherwise useless displays of plumes, ruffs, and quills—and bower birds—which assemble hoards of attractive but useless objects to impress females—show just how arbitrary the outcome of such feedback can be.

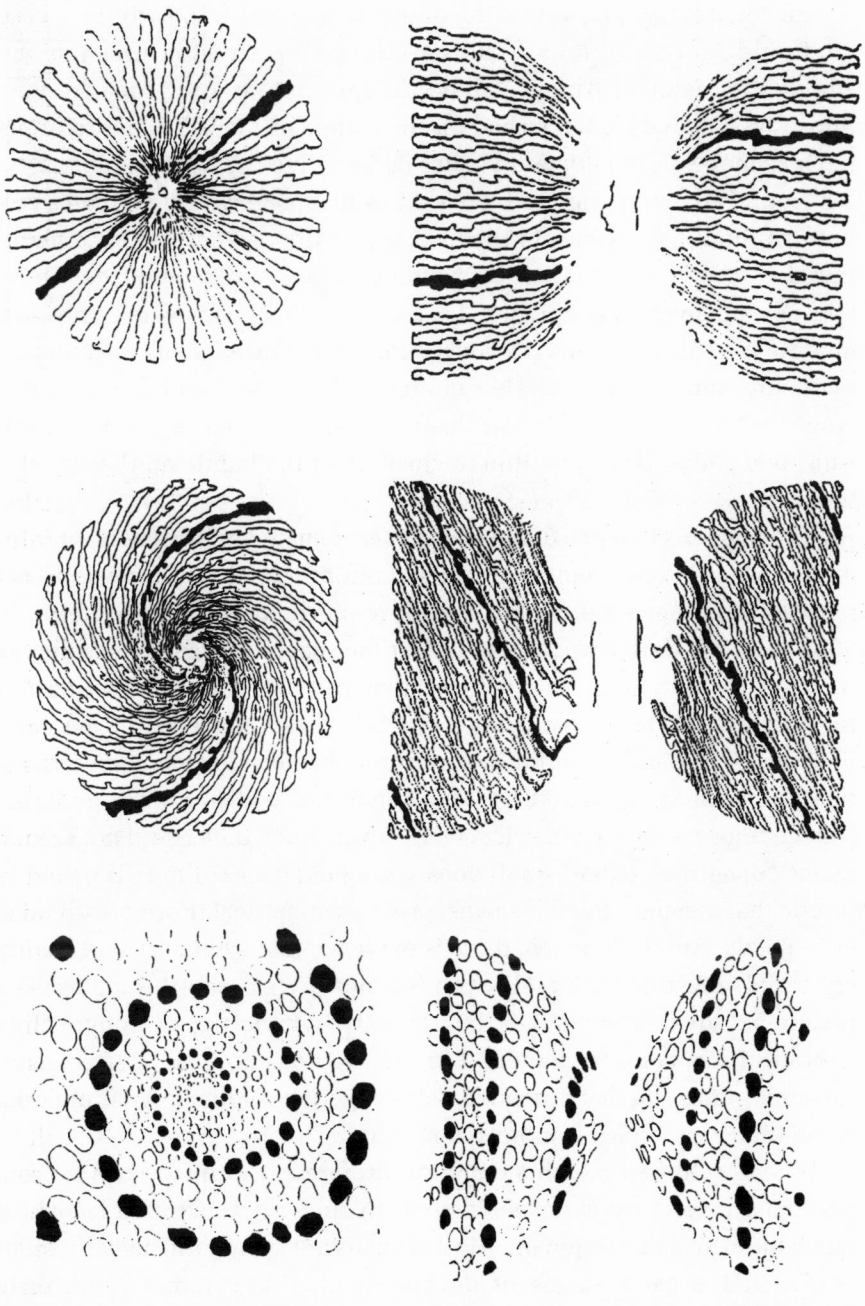

Figure 69

Common hallucinations and their transformations on the cortex, which are revealed as plane waves.

For the peacock, however, the outcome may not be so arbitrary after all. Could short, dull tails equally well have become the sine qua non among the females? Amotz Zahavi thought not, arguing that the very thing that makes a big tail so bad for the male will cause females to prefer it. Only males with very good genes can survive the handicap of carrying around an extravagant tail. So females that mate with such males will pass on those good genes to the next generation—whereas those males that do not will not, so their offspring will be at a disadvantage. Small tails never come into the picture, because *any* bird, good genes or not, can manage a small tail. Zahavi's suggestion found little favor until mathematicians such as Alan Grafen modeled the system and found that it would work, provided the handicap is a real one that really does indicate good genes. If it's possible to cheat, then the handicap theory bites the dust.

It so happens that one of the commonest female preferences is for symmetry. Many experiments and observations support the notion that females often have a preference for symmetric males. For example, in 1992, Anders Møller was trying to test the theory that female swallows prefer mates with long tails. He found that they did, but he also found that they preferred symmetrical tails, too, which was unexpected. Even more unexpectedly, Randy Thornhill discovered that female Japanese scorpion flies prefer the smell of males with symmetrical bodies, compared to the smell of those with asymmetrical ones—even when they could not see the males concerned. (Good smell goes with good genes? I'm still trying to puzzle that one out.) Bees, it seems, prefer symmetrical flowers with nice, even petals. And in humans, there is evidence that symmetry and beauty are allied, and that both males and females prefer beautiful mates. The presumed ideal human face is perfectly symmetric, although most people's faces are a bit lopsided in one way or another. There is even research indicating that human females experience more intense orgasms when their mate's face is more symmetric.[6]

Why is symmetry such a common preference? Could it be that this particular preference is *not* accidental? Symmetry is a striking feature of much biological development. Most animals are symmetric, usually bilaterally. In the early stages of development, this symmetry probably happens because the mechanisms of development are themselves symmetric, but as the embryo grows, it becomes harder and harder for symmetrically related parts of the organism to stay in step with each other. However, there are good reasons to make adult organisms symmetric—

horses wouldn't walk very well if their legs were shorter on one side than another, and birds wouldn't fly effectively with a short left wing and a long right one. So the genes probably contain a lot of instructions to ensure that development *stays* symmetric as the embryo grows. Thus, symmetry might be an effective test determining whether a genetic system is doing its job properly. Certainly, animals with defective genetic machinery are liable to exhibit asymmetries, such as a lopsided tail.

Evolution will therefore select for females that mate with symmetric males because that's one very simple way to give your offspring a better chance of inheriting those good genes. I think that the emphasis on genetics here may be overdone, so I offer one further observation: The most common way for an adult animal to be badly asymmetric is not to have poor genetics—which would make it less likely to survive into adulthood anyway—but to have been injured. In any species in which males and females cooperate to raise the children, an injured mate does not improve your offspring's survival chances.

It's a lovely tale. However, some very new discoveries suggest that sexual selection may not be the primary factor behind the preference for symmetry, after all. What else is there? Perception.

In 1994, Magnus Enquist and Anthony Arak[7] showed that preferences for symmetric mates might arise because of a side effect in the networks of nerve cells that the brain uses for perception. Such networks learn to respond to certain stimuli but not others. Imagine a network that is learning to *recognize* a tail. You have to do this before you can evolve a preference for certain kinds of tails, right? So on an evolutionary time scale, the perceptual network is subjected to various stimuli, some of them being taillike, some not. Now a tail presents itself to the eye from a variety of angles—related by the physical symmetries of the universe. If tails are lopsided, then the recognition system is likely to be presented with both left- and right-handed tails, so it will learn to respond strongly to lopsided tails in either orientation. Now think what happens when it is presented with a symmetrical tail, which bears a strong resemblance to both left- and right-handed tails. Because of the way networks of nerves respond to stimuli, the network will respond even more strongly when resemblances to *two* stimuli are present. Enquist and Arak devised computer experiments that verified this contention. At virtually the same time, Rufus Johnstone[8] carried out similar experiments and reached the identical conclusion. So a preference for symmetric mates could well be a by-product of the need to *recognize* which things are mates.

This theory does not conflict with sexual selection; it just indicates that a preference for symmetry is not an accident of fashion but is pretty much unavoidable. It also puts the human aesthetic preference for symmetry, or near symmetry, in a new light. It may be a universal, an inevitable consequence of having to evolve brains that can function in a symmetric universe. If so, then sex and beauty may not be as directly linked as has previously been thought. Even the asexual plasma-vortex aliens of Apellobetnees III will have an aesthetic preference for symmetry. Assuming they have any kind of aesthetics at all.

9

Walk on the Wild Side

All these are such as to make the whole framework act like an elastic spring, absorbing every shock as the bird lights on or rises from the ground. Bird, beast and man exhibit this resilience, each in its degree; a springy step is part of the joy of youth, and its loss is one of the first infirmities of age.

D'Arcy Thompson, *On Growth and Form,* Chapter XVI

Some years ago, I was at a mathematical conference in a British seaside town. We were staying some distance from the conference venue, and it was such a beautiful spring day that I decided to walk in. It wasn't an especially long walk, and most of it was down a single, straight stretch of road—a very gradual hill, with wide tree-lined sidewalks. The birds were singing, spring was in the air, and I was in the best of moods.

Ahead of me was a dog, a black Labrador retriever, also in the best of moods. I can see him now, in my mind, walking down the road with not a care in the world. His body was going one way, his tail another, and his legs were hitting the ground in a carefree rhythm. I'm not sure that a

Labrador can *quite* be poetry in motion—it's not the world's most elegant dog, but it's a genuinely *doggy* dog with a wonderful personality, all the same. So straight was the road, and so carefree the day, that the dog kept up the same rhythm for several minutes, undistracted by holes in the ground, bends, bumps, or any of the things that kick the brain's higher control mechanisms into action to avoid disaster: a perfect example of free-running legged locomotion in the animal kingdom. If I watched carefully, I could even see the order in which his feet hit the ground. *Back* left, *front* left, *back* right, *front* right. Equally spaced footfalls, following the same pattern, over and over again.

It really is a pattern. Look at the words, say them to yourself in your mind. Tap the table as you do so. Back, front, back, front. Left, left, right, right. Two intertwined mathematical sequences that capture the essence of the doggy walk—and the walk of an ox, a horse, or an elephant . . . indeed of any four-legged animal that walks.

Only as I write this introduction does it occur to me that I, too, was walking without a care down a long, straight road—and I was a perfect example of two-legged locomotion in the animal kingdom, but my pattern was simpler: left, right. The patterns with which animals move their legs—known in the trade as "gaits"—have long intrigued zoologists. There are all sorts of reasons for wanting to understand animal gaits. One reason is that such understanding can help us detect problems with such things as artificial hip joints: One of the signs that a person's artificial hip is coming adrift is a change in the way the person walks. Another reason, very much in tune with the modern age, is that animal locomotion offers useful hints to the designers of legged robots. There are many places where a human cannot safely go, and where a wheeled robot can get stuck: military target ranges for testing weapons, defunct nuclear reactors that have to be decommissioned, or even the surface of Mars. Terrestrial evolution long ago discovered most of the tricks for making legged creatures move stably and efficiently, and many robotics engineers take their cues from animal movement. After all, there's no point in reinventing the leg.

The constraints on an organism's movement are those of physics. If it uses limbs, they must be strong enough to withstand the forces that act on them. (I have seen badly designed robots tear themselves to bits when they move.) The same goes for other forms of locomotion. If the animal swims, then it has the laws of fluid mechanics to contend with. It is hardly surprising that the influence of the laws of physics is very noticeable when it comes to animal movement. Very clearly, this is a case where mathematics has provided an extensive catalog of patterns, and biology has plundered it. Scarcely a possibility goes unused, however exotic.

The influence of physics goes deeper, too. It's no use having legs unless you have a nervous system to control them. Locomotion and neural nets go hand in hand, and they must have evolved together, not one at a time. Also, just as the neural nets for sensation must mimic the patterns of the outside world, so the neural nets for locomotion must mimic the mechanical patterns of animal bodies. I have a strong suspicion that this coevolution was made possible—or at the very least far easier —because of a remarkable fact: The natural patterns of oscillation for mechanical systems such as limbs are the same as those for neural nets. There is a universality to gait rhythms, which potentially linked animals' limbs to their brains long before limbs or brains existed as fully formed biological structures. Gait rhythms offered a pattern sitting in evolutionary phase space, waiting to be used.

And used it has been. Virtually all living creatures move. Even the most sedentary plant bends toward the light, and even the tiniest speck of plankton drifts on the ocean currents—but the cheetah can chase its prey at up to 110 kilometers (about 68 miles) per hour. That's *really* moving.

There are almost as many ways to get about as there are organisms. Bacteria propel themselves through water using tiny rotating helical screws, like a ship. Single-celled organisms such as *Paramecium* can choose their direction of motion by waving their whiplike cilia. The mathematical patterns in locomotion are striking: The cilia of the paramecium move in traveling waves, like a field of corn in a breeze, and nothing could be more geometric than the bacterium's rotating helix. Snakes and eels wriggle their way through space on waves of muscular contraction (Figure 70). The sidewinder, a kind of snake, rolls its way across the hot desert like a coiled spring. The inchworm walks its tail end up to its head, while its body forms a ∩ shape; then walks its front end away again to straighten out to a —. The albatross glides on rigid wings, with the occasional lazy flap, and clumps its way across the water on webbed feet in a charmingly clumsy takeoff run. The elephant ambles ponderously across the savannah, moving one foot at a time (Figure 71) in the same pattern as the walking dog in the seaside town. The camel uses a different pattern: It moves both left feet at once, then both right, and sways from side to side like a drunkard. The squirrel employs another pattern: It hops, pauses, hops again; If alarmed, it dispenses with the pauses. The wheel

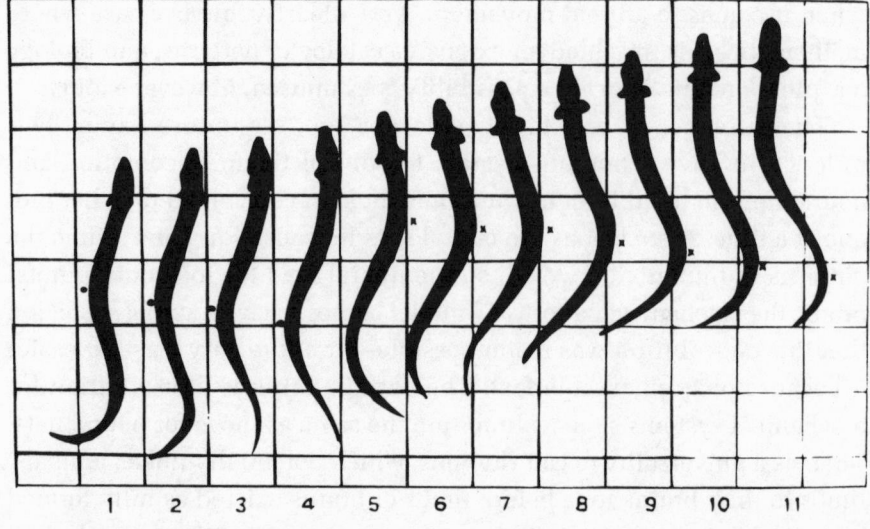

Figure 70
Waves of muscular contraction in the eel *Centronotus gunnelus*.

spider *Caparachne aureoflava* rolls across the desert[1] like an eight-spoked wheel. And the world's only jumping maggot, more soberly known as the fruit fly larva *Ceratitis capitata*, contorts itself into a U-shape and then straightens out (Figure 72), leaping into the air and following a perfect parabolic arc, like a cannonball in a medieval ballistics text.[2]

Figure 71
The amble of the elephant.

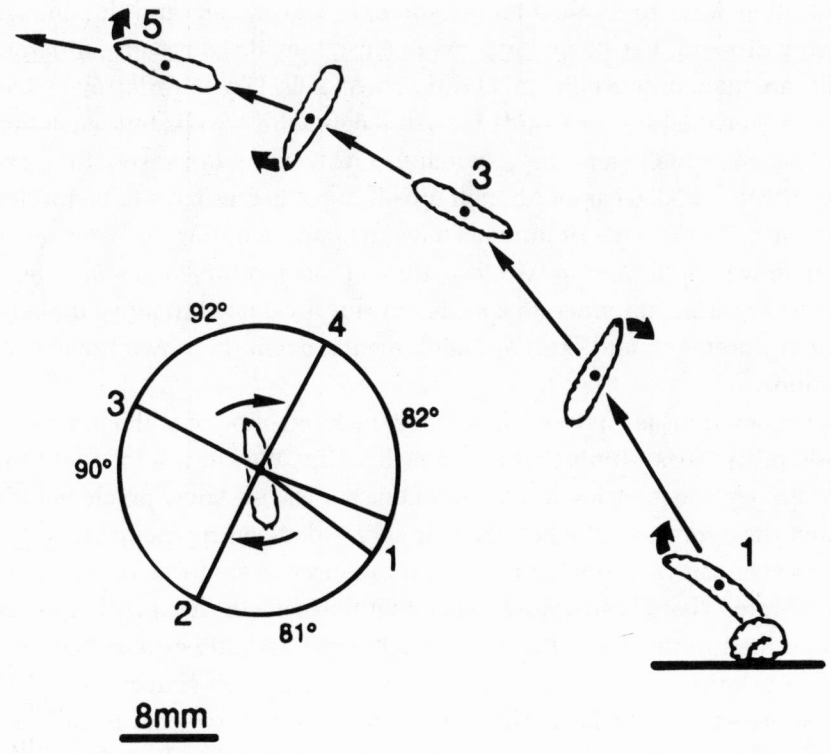

Figure 72
The jumping maggot.

There are endless mathematical patterns in locomotion, and for once, this is an area where genetics seems to have little (yet) to tell us. It is the mechanics of locomotion that matters—or, more properly, the *mechatronics* (*mechanics* plus *electronics*). This term was introduced in robotics, to refer to human-made machines, but it captures the essentials of animal movement: electrical activity in neural nets to create and control the underlying rhythms, coupled to muscles and bones and *chitin* (material composing insects' exteriors) to create the actual motion.

It would be easy to devote the rest of this book to locomotion, so to focus the discussion, I concentrate on legged animals—though the legless lamprey, an eellike creature that attaches itself to rocks, will feature later in this chapter, too, for good reasons. What makes the mathematics of locomotion so fascinating is that nature employs a huge variety of different patterns.[3] Many animals use only one gait (aside from stand, sit, and other ways to nonlocomote); some, however, can (and do) change their gait—and, like changing gears in a car, they do so because different gaits are more or less efficient at different speeds. (Try to *walk* 100 meters [about 90 yards] in 10 seconds.) Horses begin with a walk, but when that gait becomes inefficient, they switch to a trot, and at top speed, they gallop. Some can also canter, though not all; most horses have to be trained to canter, just as pairs of humans have to be trained to waltz. Also, even though we are limited to two legs, human locomotion comes in several kinds: We walk at moderate speeds but run if we need to move quickly. Our children hop and skip; our adolescents invent their own novel contortions.

Locomotion, as I have said, requires the interaction of both neural nets and bodily parts—limbs, wings, muscles, fins. We know a lot about the physiology, the muscles and bones of our bodies; we know precious little about the neurology, the networks of nerves that control locomotion. We don't even know where the neural control circuits are located, though in vertebrates, there is strong evidence that they live in the spinal column, rather than in the brain. It is believed, but not yet fully established, that rather primitive neural nets called "central pattern generators"—"CPGs" for short—set up the basic rhythms of *free locomotion*—that is, locomotion unimpeded by variations in terrain, like that of the black dog on the long, straight slope.

The problem is that we cannot—yet—dissect out specific parts of a complex nervous system or understand exactly what those parts do. Nonetheless, we are now able to make some educated guesses as to what a CPG circuit looks like—guesses that are good enough, for example, to suggest new ways to design walking robots. Next, I tell you the story of what these guesses are, and where they came from. Mathematics has been indispensable; so, of course, has experimental biology. Genetics, on the whole, has played a very minor role, probably because few people seem terribly interested in the genetic basis of locomotion. That interest, I am sure, will increase.

A basic mathematical feature of gaits is *periodicity:* Unless the animal is affected by variations in terrain, the presence of another animal, or some other extraneous influence, or unless it wishes to change speed, it will repeat the same rhythmic movements over and over again. Another key mathematical feature is symmetry, and the ubiquity of symmetry in gaits was emphasized in 1965 by the American zoologist Milton Hildebrand.[4] For example, when an animal bounds, both front legs move

Figure 73
The bound of the long-tailed Siberian souslik.

together, and both back legs move together (Figure 73). This motion is (among other things) symmetric under left–right reflection of the animal. Other gait symmetries are more subtle: for example, when a camel paces, its left half moves in the same manner as the right half, but half a period out of phase—that is, after a time delay equal to half the gait period (Figure 74). This is an example of *spatiotemporal symmetry*, combining change in both space and time. Why do gaits have spatiotemporal patterns? The answer seems to be related to the mathematics of *oscillators* (things that change periodically). There is a striking analogy between animal gaits and universal periodic patterns in simple networks of oscillators. This analogy suggests that the patterns are natural consequences of the animal's physiology or neural circuitry, and it provides clues about how its neural control circuits might be organized.

A gait's qualitative form is captured by a list of the relative phases of the feet. The relative phase of a given foot is the fraction of the gait cycle between a reference foot hitting the ground, and the foot concerned hitting the ground. For example, in the human walk, the right foot hits the ground half a period later than the left, so if we take the left foot as reference foot, then the right foot has a relative phase of 0.5. The reference foot always has a relative phase of 0. There are two basic bipedal gaits. The legs may be out of phase with each other, generally by half a period— that is, doing the same thing but hitting the ground half a gait cycle apart (walking and running are examples in humans). Alternatively, the legs can be in phase, with both of them doing the same thing at the same time (for instance, two-legged hopping). There are other human gaits, such as the skip, but these are secondary phenomena.

Quadruped gaits have more variety. The phase relationships for the eight commonest quadrupedal gaits are summarized in Figure 75.

Different animals prefer different gaits. Most walk at low speeds. As the speed gradually increases, the trot is the next choice for most four-legged mammals, although camels generally pace. Wildebeests change directly from a walk to a canter. Reptiles can trot, and some lizards can run on their hind legs. Young crocodiles, but not adults, can gallop, both rotarily and transversely.

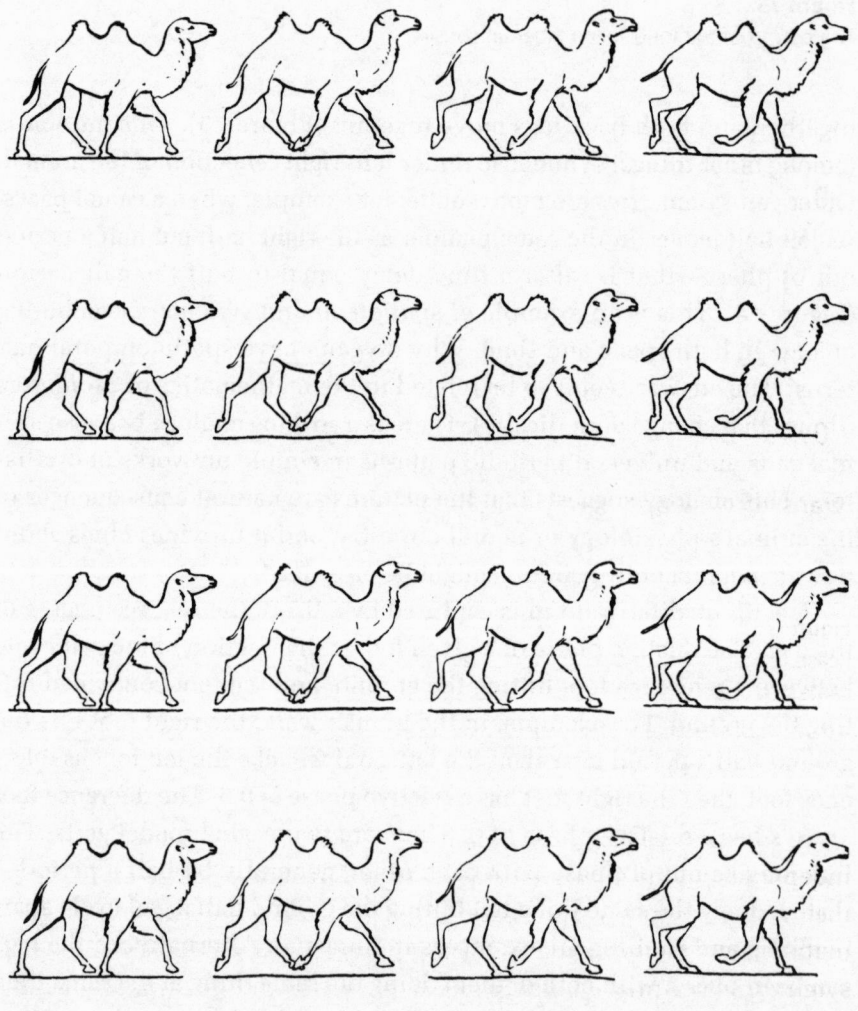

Figure 74
The pace of the camel.

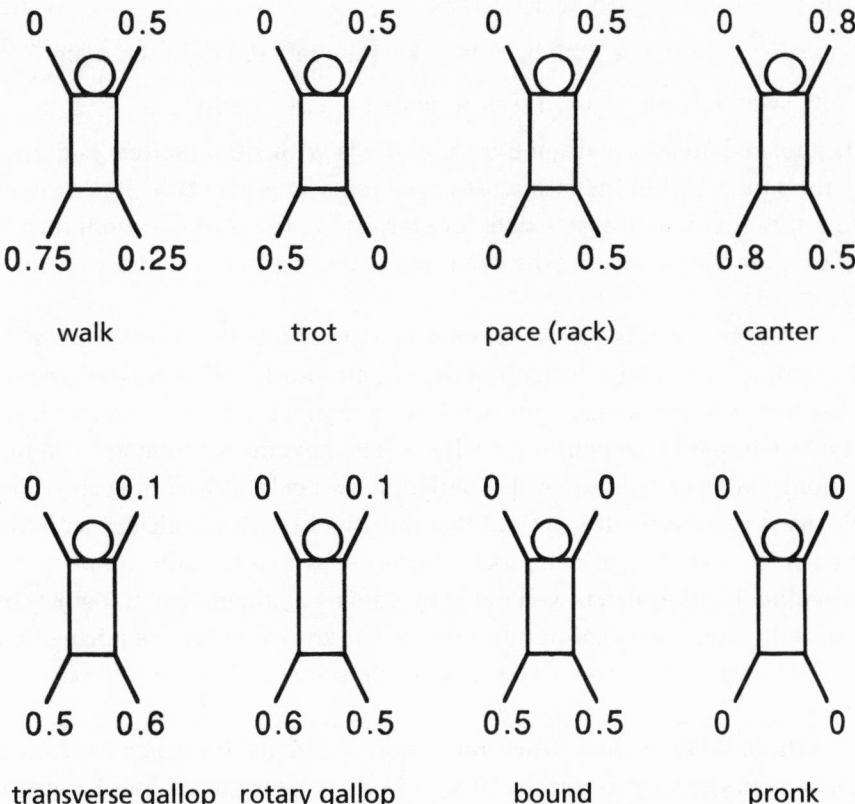

Figure 75
Phase relationships for the eight most common quadraped gaits: walk, trot, pace, canter, transverse gallop, rotary gallop, bound, and pronk.

In the early 1990s, G. N. Schöner, W. Y. Yiang, and Scott Kelso—and independently Jim Collins and I—observed that gaits can be classified by their spatiotemporal symmetries.[5] *Spatial symmetries* are (notional) permutations of the legs, such as "interchange front and back." *Temporal symmetries* are phase shifts. For example, the quadrupedal bound has two fundamental symmetries:

◇ Swap left legs with right legs, keeping front and back the same.

◇ Swap front and back *and* shift phase by half a period.

The pace has a slightly different two:

◊ Swap front legs with back legs, keeping left and right the same.

◊ Swap left and right *and* shift phase by half a period.

The formal area of mathematics that deals with such matters is called "group theory," but the formalities need not concern us here. It is enough to know that machinery exists for classifying the possible symmetries of gaits, and for analyzing the dynamics that can give rise to these symmetries.

In these terms, the most symmetric gait of all is the *pronk,* in which the animal repeatedly leaps into the air, all four feet leaving the ground together—though we can question how common the pure pronk, with all the legs in perfect synchrony, really is. The next most symmetric are the bound, the pace, and the trot; finally, the two gallops and the canter are the least symmetric. It turns out that this hierarchy is consistent with the dynamics of oscillator networks. The more symmetric gaits are *primary,* meaning that they can be generated by a single mathematical transition in the gait pattern. The gallops and the canter are *secondary,* requiring two such changes. In a sense, the secondary gaits are combinations of primary ones.

Where do the symmetries come from? Animals' brains control their movements by sending electrical signals along neurons. As I said, there is strong circumstantial evidence that the basic patterns of locomotion in an animal are created by a CPG: a neural circuit—not necessarily in the brain—that produces rhythmic behavior. Rhythms are bread and butter to neural nets, just as they are to electronic circuits; in fact, the mathematics of nerves and electronic circuitry is extremely similar, and for many purposes, the same models work for both. One of the most basic electronic circuits, as every engineer knows, is the *oscillator,* a circuit creating a voltage that switches from positive to negative and back again a huge number of times per second. Some oscillators generate radio and TV signals, and other oscillators modulate the signals to impress a recognizable signal into their waveform.

What's needed for a CPG is not just one oscillator, but a whole system of oscillators—a separate oscillator for each leg, at least. The idea is that the legs—themselves oscillators, though mechanical ones, not much more than complicated pendulums—swing in patterns of movement that are orchestrated by the oscillators in the CPG. Whatever rhythm the CPG puts out, that's what the legs do. The separate oscillators must be wired together in the CPG, forming a kind of circuit, so that each knows what the others are doing. CPG oscillators have to be a lot slower than those in a radio transmitter because legs don't move thousands of times every second. Also, CPG circuits have to be quite versatile because legs must be able to move in a variety of different patterns, to select the best gait for a given speed.

Trying to locate a specific piece of neural circuitry in an animal's body is rather harder than searching for a needle in a haystack the size of the Great Pyramid, so despite a widespread belief that CPGs exist, the direct evidence is slight, except for a few animals such as the lamprey. Most attempts to deduce the circuit diagrams of CPGs therefore start from the observed gaits and try to work out what sort of circuit might produce them. However, some very impressive work on the swimming movements of lampreys, an eellike creature, have worked directly from clues about the animal's nervous system and then deduced how the creature moves, as I describe at the end of this chapter. Perhaps the most interesting thing is that both approaches seem to be converging on the *same* general circuit design.

First, I begin with the simplest interesting case: bipeds. The simplest plausible network for a biped has two identical components, one for each leg. The components themselves might be quite complex neural circuits, but we're not interested in their internal structure: All we need to know is that for both legs, *they are both the same.* Because we can't dissect them out, we can't be certain of this, but the bilateral symmetry of animals and various other considerations make it a good bet. So powerful is the mathematical machinery of oscillator networks that this simple assumption can be shown to imply that for *all* such CPGs, there will be two primary oscillation patterns:

◇ The in-phase pattern, in which both oscillators have the same waveform

◇ The out-of-phase pattern, in which both oscillators have the same waveform *except* for a half-period phase difference

These two types of signals produce exactly the two primary bipedal gaits. (The skip arises from the walk, by way of a secondary phenomenon called "period doubling"; I mention this because at least one gait analyst is convinced that I don't know that children skip.)

What about quadrupeds? The most natural assumption is that their CPGs involve four oscillators. This is what Collins and I assumed—until the late 1990s—and we showed that various networks of four oscillators can reproduce, among them, all the common gaits, except perhaps the canter. However, there were some technical snags, the most serious being that any network that can generate both a trot and a walk will also generate a pace, under *identical* conditions to the trot. This sits uneasily with the fact that camels pace but never trot, while most horses trot but never pace.[6] There was another snag, too: Our network for "walk" also produced some distinctly curious gaits that didn't seem to occur in real animals.

We weren't sure how to get around these problems, or even how serious they were, so we moved on to the next case: six legs—insects, mainly. Interest in insect gaits is centuries old. In the early 1600s, Galileo used his newfangled microscope to discover that these multilegged animals can move in antigravitational conditions. Some insects can hang upside down and walk along the surface of an object by using the sucker pads or small claws located on their legs. Somewhat later, Giovanni Borelli, in his *De Motu Animalium* of 1680, argued that during walking, the three legs on each side of an insect must move in the order rear, middle, front. Interestingly, Borelli's prediction was not based on observation. Instead, it was founded on the theoretical principle that the back leg should move first in order to propel the insect's body forward before the other legs are lifted off the ground. As it happens, Borelli was more or less right, though it's not clear that he deserved to be.

The most common insect gait is the *tripod* (Figure 76a). In this pattern, the back and front legs on one side of the animal move in phase with the middle leg on the other side, and then the other three legs move together, with a relative phase of half a cycle compared with the first set. The alternating sets of three legs form triangular bases of support on the ground—hence, the gait's name. This pattern is typically used at medium and fast speeds. At slower speeds, insects often adopt the metachronal-wave gait (Figure 76b). In this pattern, a wave of leg movements sweeps from the back of the animal to the front, first on one side then on the other. Adjacent legs on one side of the insect have relative phases one sixth of a cycle apart, and the two legs on each segment have a relative phase of half a cycle. Although these two gaits are the most typical, there are a number of other possibilities. In fact, insects are not limited to six-legged gaits. In the mid-1990s, Robert Full and Michael Tu of the University of California at Berkeley found that the American cockroach switches to a four-legged or even a two-legged gait when it needs to run fast.

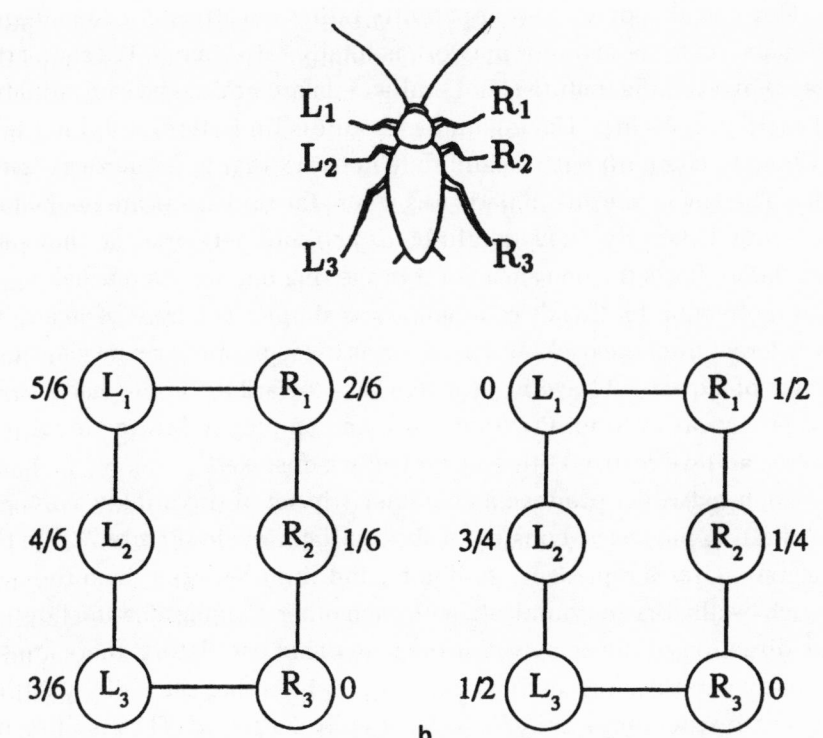

Figure 76
Two common insect gaits: (a) the tripod, (b) the metachronal wave.

By adapting our ideas on quadrupeds, Collins and I found that a simple network, comprising six identical oscillators linked in a ring, produces many of the classic hexapod gaits.[7] This was encouraging, but the quadruped theory still suffered from some technical problems, some of which also caused difficulties with our theory of hexapods. In late 1996, during the final stages of preparing this book, I spent a month in Houston working with Marty Golubitsky, an American mathematician who has long been a close friend. He convinced me that those problems were significant, and that a radical redesign of the network was in order. We spent several days looking for a universal network that would produce the right patterns for quadrupeds and hexapods—and even centipedes if need be— one that produced *all* the required gaits, preferably *no* unwanted ones, and kept trot and pace genuinely different.

Our first discovery was apparently rather negative: For four-legged animals, *no* four-oscillator network is totally satisfactory. We soon realized, however, that nature is not limited to maintaining the same number of oscillators as legs. Having understood this limitation, it did not take us long to come up with a candidate network that is far more satisfactory. The key is to think of networks where the oscillators are hooked up in a ring. Crucially, they should be *directional* networks, so that each oscillator affects the one ahead of it in the ring but not the one behind it. Our motivation for this directionality was simple: The front of an animal is different from the back, and most forms of locomotion can be viewed as waves of movement traveling from the back toward the front—hardly ever the other way around. Previous work had employed bidirectional networks, and we realized this was probably a mistake.

Such networks produce a characteristic set of traveling waves. For illustrative purposes, consider a six-oscillator cycle (Figure 77). In the diagram, circles represent oscillators, and lines between them indicate which oscillators communicate with each other. Communication flows in the direction of the arrows. For the ring of six oscillators, there are six primary patterns. Successive phases around the ring can shift by 0 (that is, stay in synchrony), or by $\frac{1}{6}$, $\frac{2}{6}$, $\frac{3}{6}$, $\frac{4}{6}$, or $\frac{5}{6}$ of a period. The resulting list of patterns has a very neat arithmetical structure, which looks like this:

> *Pattern 1:* 0, 0, 0, 0, 0, 0
> *Pattern 2:* 0, $\frac{1}{6}$, $\frac{1}{3}$, $\frac{1}{2}$, $\frac{2}{3}$, $\frac{5}{6}$
> *Pattern 3:* 0, $\frac{1}{3}$, $\frac{2}{3}$, 0, $\frac{1}{3}$, $\frac{2}{3}$
> *Pattern 4:* 0, $\frac{1}{2}$, 0, $\frac{1}{2}$, 0, $\frac{1}{2}$
> *Pattern 5:* 0, $\frac{2}{3}$, $\frac{4}{3}$, 0, $\frac{2}{3}$, $\frac{4}{3}$
> *Pattern 6:* 0, $\frac{5}{6}$, $\frac{2}{3}$, $\frac{1}{2}$, $\frac{1}{3}$, $\frac{1}{6}$

The way to list the patterns is to choose a phase, say $\frac{1}{6}$, and then to write down what you get when you multiply that phase by 0, 1, 2, 3, 4, and 5. Bear in mind that a phase of 1 is the same as 0. For the complete list, you apply this calculation to all possible phases: 0, $\frac{1}{6}$, $\frac{1}{3}$, $\frac{1}{2}$, $\frac{2}{3}$, $\frac{5}{6}$. That's it.

All of these patterns (except the first) represent traveling waves. For instance, in Pattern 2, with phases 0, $\frac{1}{6}$, $\frac{1}{3}$, $\frac{1}{2}$, $\frac{2}{3}$, $\frac{5}{6}$ around the ring, the wave starts at oscillator 1 and moves on in turn to 2, 3, 4, 5, and 6. In contrast, in Pattern 3, the phases are 0, $\frac{1}{3}$, $\frac{2}{3}$, 0, $\frac{1}{3}$, $\frac{2}{3}$, so oscillators 1 and 4 move together, then 2 and 5, then 3 and 6; and so on. Pattern 4 groups

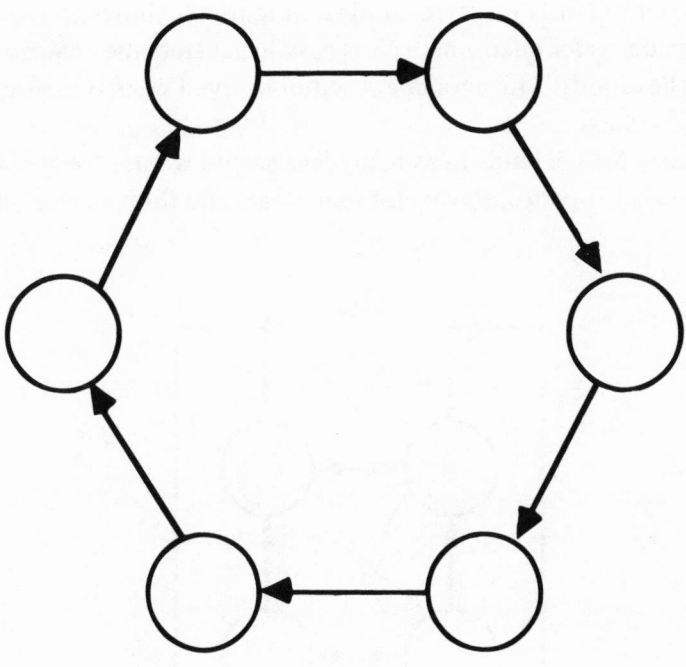

Figure 77
A cycle of six oscillators: The oscillators are represented by circles, and the connections between them by arrows.

the legs in two sets of three, half a period out of phase—just like the tripod gait. That's encouraging because most animal gaits are traveling waves from front to back, but often with a phase difference of ½ from left to right. (If not, left and right sides are usually in phase.) This suggests using two rings of oscillators, one for the left legs, one for the right. It turns out that for animals with four legs, the choice that mimics the observed gaits best uses *eight* oscillators—two rings of four, cross-connected left to right (Figure 78). Only four of these oscillators actually drive the legs; the other four are used to propagate the traveling waves from the front of the animal to the rear, to maintain the correct timing and phase relations. Similarly for insects—hexapods, six-legged beasts—you need *twelve* oscillators, two rings of six. And so on.[8]

So that's what we were missing: an extra, hidden set of oscillators.

Now everything falls neatly into place. Figure 79 shows the resulting primary patterns for quadrupeds; it is easy to generate the analogous diagram for hexapods. The agreement with observed gaits turns out to be gratifyingly close.

The same idea extends to as many legs as you want. *Myriapods* (centipedes and millipedes) have a lot more legs, and they mostly produce

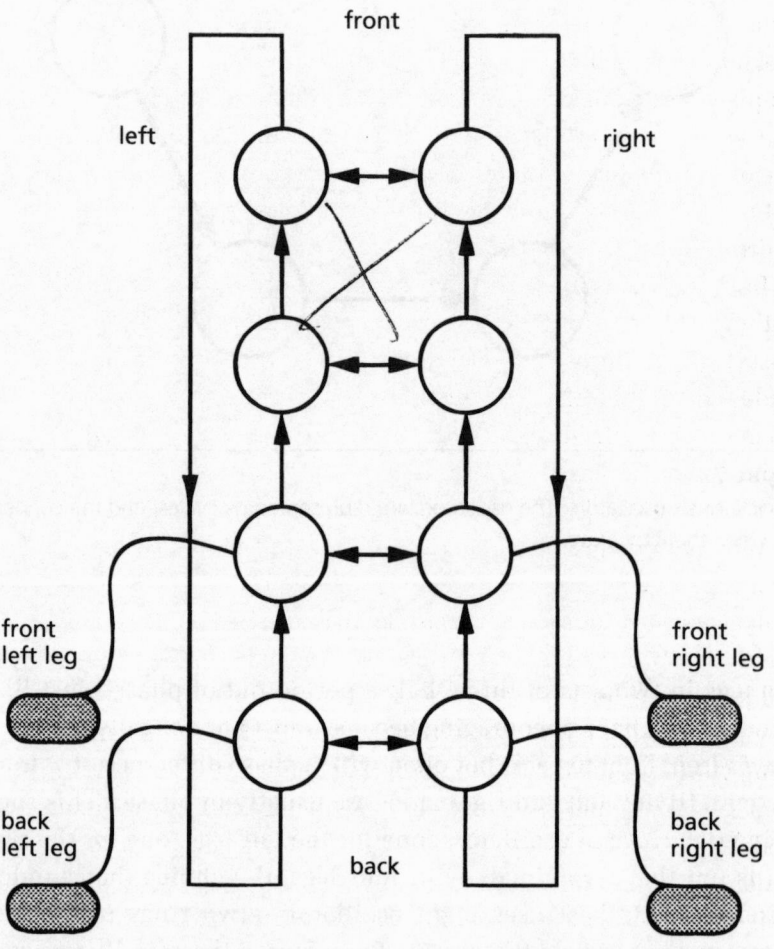

Figure 78

Two cycles, each with four oscillators, can model quadruped gaits. One cycle controls the left legs, the other controls the right legs. Only four oscillators are directly connected to legs; the others form a return circuit for the signals.

rippling patterns of leg movements (Figure 80). These patterns, too, can be produced by exactly the same network design, again with twice as many oscillators as legs. That design, by the way, is schematic: Any network with the same symmetries will produce the same range of gaits. As I write, Luciano Buono has found that adding extra connections, symmetrically arranged, helps to produce stable gaits more easily.

When we consider possible secondary patterns, there's an unexpected bonus: The quadrupedal network can also exhibit motions with the same patterns as the two gallops—*and* the canter, which we'd never been able to obtain convincingly before. These gaits arise as mixed modes, in which the phase relationships of two distinct primary gaits are combined. The *transverse gallop* is a mixture of trot and bound, the *rotary gallop* a mixture of trot and pace, and the *canter* is a mixture of walk and jump, a new gait rather like a bound but with a brief pause—just like a slow-moving squirrel.

In other words, we've shown that there are universal patterns in the mathematics of oscillator networks, and that those patterns match, very closely, the patterns found in animal gaits. It seems reasonable to conclude that animal CPG designs use those patterns. The prediction is that the neural circuitry in animal CPGs should correspond, in general terms, to our schematic mathematical networks. Specifically, they should have the same symmetries.

Whether this prediction is right must await more detailed knowledge of CPGs, although we can hope for indirect confirmation by pushing the analysis farther, to look at transitions between one gait and another. That ought to produce some further predictions, which can be tested against real animals. Incidentally, it is only necessary to *observe* the animals moving: Nothing harmful need be done to them. Moreover, 1990s work along similar lines has led to some predictions about biology that have been verified experimentally—a major success for oscillator-network models. These predictions have been obtained by Nancy Kopell and Bard Ermentrout,[9] and refer to the lamprey CPG. Perhaps not coincidentally, their network has a lot in common with ours, although the way they invented it was quite different. The lamprey has no legs; instead, its movement is produced by waves of contraction in pairs of muscles, spaced much like the legs of a centipede. Abstractly, it's the same kind of system as the quadripedal one, and similar ideas apply. The work of Kopell and Ermentrout is directly related to the animal's known physiology.

Prior to Kopell and Ermentrout's work, in 1982, Avis Cohen, Philip

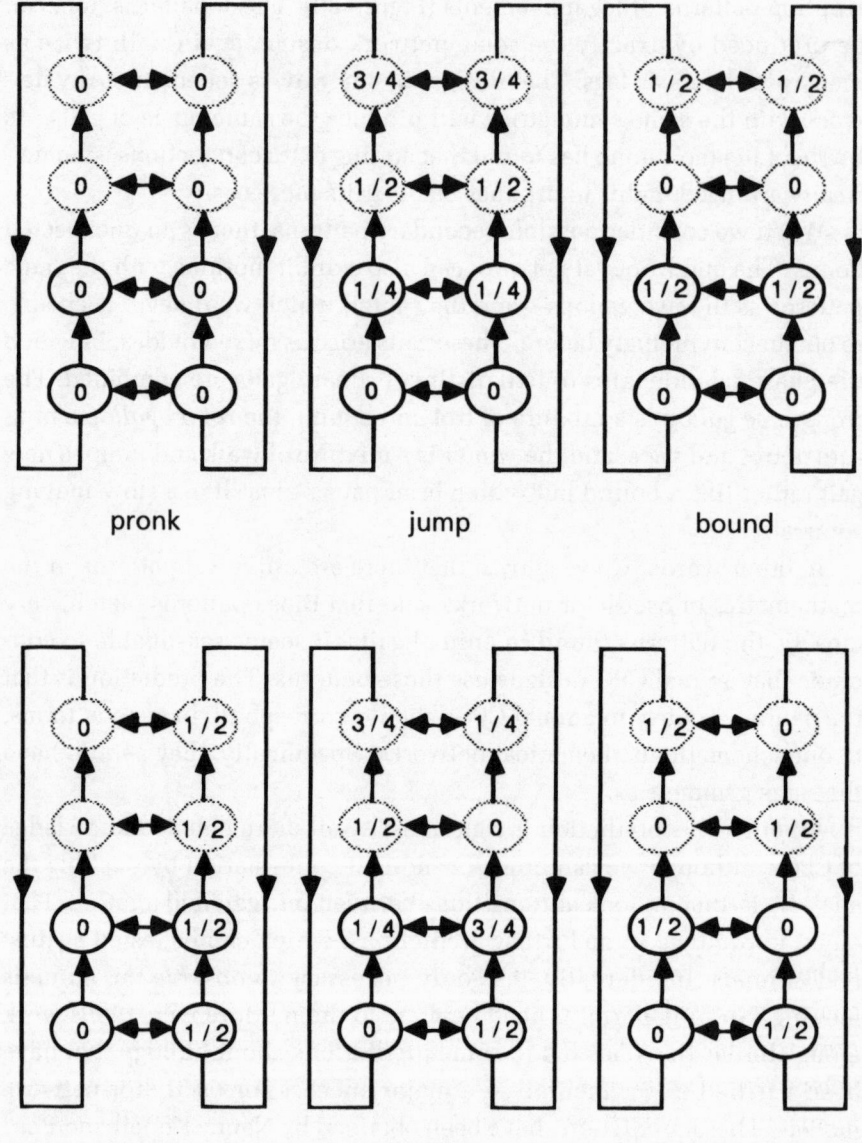

Figure 79
Primary gait patterns predicted by the eight-oscillator network for quadrupeds, labeled with the corresponding animal gaits. Numbers indicate phase shifts.

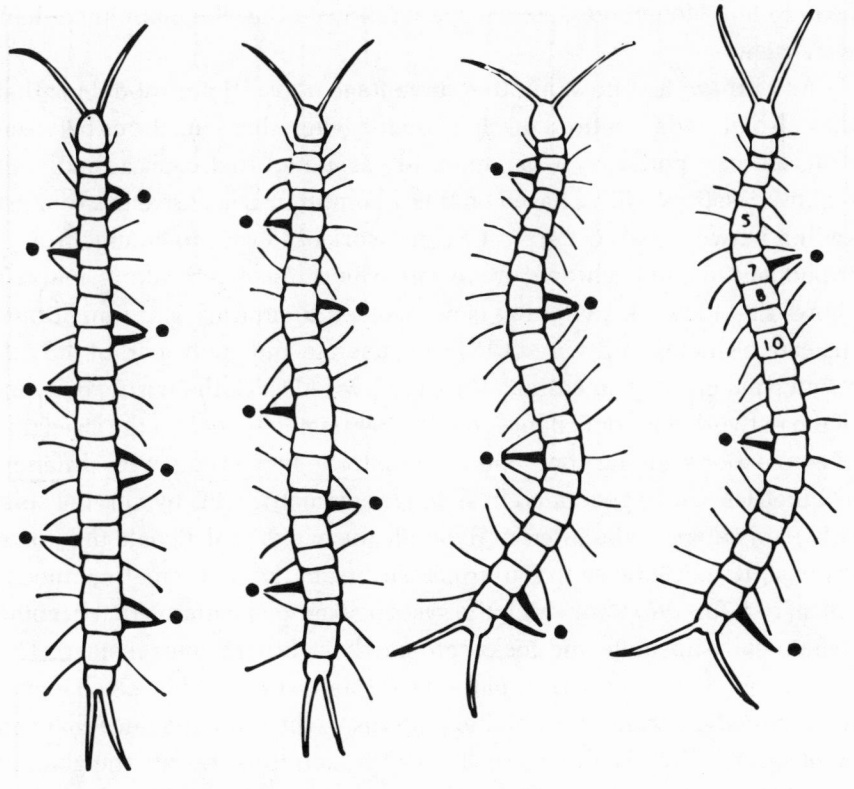

Figure 80
Rippling waves in the centipede.

Holmes, and Richard Rand had modeled the lamprey spinal cord using a chain of coupled oscillators.[10] Kopell and Ermentrout continued the lamprey work, developing their own model based on *synaptic coupling,* where two oscillators influence each other, even if they are behaving identically. As with the quadripedal network discussed previously, their synaptic coupling is biased in one direction, and the network generates traveling waves. The two researchers predicted how the network would behave if one end was *forced*—subjected to an incoming periodic pulse—and their predictions were verified in actual lamprey spinal cord by biologists Karen Sigvardt and Thelma Williams. The most striking prediction by Kopell and Ermentrout came a little later: They showed that in order to generate the required waves of movement, tail-to-head coupling must be stronger than head-to-tail coupling. Biologists simply hadn't expected this because in normal swimming, the wave goes the other way, from head to tail. Nonetheless, experiments showed that the mathematicians were right.

As I hinted just now, another advantage of oscillator models is that they also provide a natural mechanism for generating transitions between different gait patterns. It is commonly assumed that each animal gait requires a separate CPG. Based on this assumption, a cockroach, say, must switch between two different CPG networks in order to change from a tripod gait to a metachronal-wave gait. The symmetry-breaking analysis shows that network switching is not needed to produce gait transitions. Instead, a single, hard-wired CPG can generate and control all of the different gaits used by an animal. (In a hard-wired CPG, the wiring diagram of the network does not change, but the parameters—such as the strengths of connections among component oscillators—may.) Gaits with different symmetries can be generated by a *single* symmetric CPG by varying suitable parameters in the internal dynamic, or in external signals that force it to oscillate. Changes in patterns result from the mathematical mechanism of *bifurcation,* in which the system abandons patterns that become dynamically unstable and locks into newly stabilized ones instead. The parameters determine which patterns are stable or unstable, so changing the parameters changes the stable patterns—which are the ones that can be observed directly in nature. This approach thus reveals the general oscillation patterns, and the transitions between them, that should be expected of a CPG model—or even whole classes of CPG models—independently of specific aspects of the oscillators' intrinsic dynamics or the nature of the coupling. The general form of the results depends *only* on the symmetry of the network. In fact, such a CPG can be made to switch between different gait patterns, merely by varying the nature of its driving signal, which could originate in the brain.

Not only is this scheme simpler than a gearbox scheme with several independent CPGs, but it also provides a far more plausible route for the evolution of multiple gaits. What possible competitive advantage could a newly evolved rudimentary CPG for trotting offer to an animal that already had a sophisticated CPG for walking? How, then, could a gearbox design evolve? On the other hand, if trotting automatically comes along with walking in the same CPG, the two gaits can coevolve in an entirely natural way, along with the entire system for switching between them.

So, somewhere in the nervous system of that black Labrador, there is probably a simple network of neurons that generates all of the underlying rhythms with which he moves. His legs, and the neurons that control them, are utilizing a pattern as old as time itself, one that goes right back to when the very first neural nets evolved in living creatures—the rhythmic patterns of oscillator networks. And we can tell that such a network must be there without in any way disturbing his happiness as he waddles his carefree, doggy way down that long seaside hill.

10

An Exaltation of Boids

[A] flock of flamingos, wearing on rosy breast and crimson wings a garment of invisibility, fades away into the sky at dawn or sunset like a cloud incarnadine.

D'Arcy Thompson, *On Growth and Form*, Chapter V

In January 1996, I was visiting friends in Monte Carlo, and I witnessed one of the most extraordinary sights I have ever seen. From high up in one of the innumerable tower blocks that climb skyward from the tiny principality's precious ground, I was looking down into the harbor, jam-packed with expensive yachts, glowing in the light of the setting sun. Way down below me, in among the high-rises, was a huge tree—a tree laden not with fruit, but with birds. Literally thousands of birds—they looked like starlings but I can't be sure—were spiraling in to settle down for the night. So great was the mass of birds that a bird-scarer, employed specifically for the purpose, shortly appeared to chase them away, before they stripped the branches off the tree.

The sheer concentration of birds was amazing enough—but there was more. As the birds dipped and wheeled in the sky, the entire flock slowly began to acquire some kind of collective order. At first, it seemed little more than a random swarm of independent birds, a vast fuzzy scattering of black dots in the sky, but as the sun dipped lower, the entire flock slowly began to act as one—like some gigantic, ethereal skyborne organism. The flock acquired sharp edges, a dense black swarm with its own collective will. The swarm spiraled through the sky, swerving and turning with astonishing speed and an impressive unity, as if each bird knew exactly what to do. Occasionally, a large clump would split off, whirling and swirling in its own independent dance—yet soon it rejoined the main group, as if drawn by an invisible magnet. A few stragglers flapped aimlessly about, apparently seeking to join the main flock but unsure of the best route; the majority of the birds behaved as one. Then, magically, the flock dispersed: The dance was over.

It happens, I was told, every winter evening.

There is something utterly awe-inspiring about large groups of animals—especially their apparent unity of purpose. We wonder where it comes from, how the individuals know what the group is supposed to do, and how they play their part in achieving it. It is not just flocks of birds (Plate 16) that exhibit striking patterns of collective behavior. Schools of fish create glittering swirls of movement in tropical oceans, flashing this way and that, but never leaving the group—stopping and starting in an instant (Plate 17). Huge herds of wildebeests trek for hundreds of kilometers across the African veldt, following the tracks of ancient migrations, millions of years old. Not so long ago, equally vast herds of bison roamed the plains of America. In the rain forests, long trains of ants ferry food or building materials into the nest. In some parts of the world, termites construct weird vanes, deep below ground, to equip their nests with air conditioning.

What can possibly be responsible for the remarkable behavior of social animals? What gives them the appearance of possessing a group mind, as if some central conductor were orchestrating their behavior? Catchall terms such as *instinct* only deepen the puzzle: They surely do not solve it, for what *is* instinct? How come a humble termite is so massively endowed with instinct that a group of them knows how to install air conditioning in the nest? It's not instinct. It's rules. Over hundreds of millions of years, evolution has exploited patterns of collective behavior that arise, spontaneously, from the abstract mathematical rules that physics provides. It has built those patterns into the animals' genes and their social behavior—some of which may be passed on through learning, not genetics, at least in the higher animals. Evolution has not built the patterns in *directly*, however. There is—I strongly suspect—no genetic instruction "form a flock" in a bird. Instead, there are genetic and behavioral analogues of the *rules* that produce flocks. Evolution has constrained the repertoire of bird behavior, both genetically and culturally, to incorporate such rules.

Why do I think evolution has favored the rules themselves and not their consequences? There are four reasons. The first is efficiency: On the whole, the rules are simpler than the behaviors they generate. A few rules, with built-in contingency planning, can encapsulate a *huge* range of behavior, adaptable to a huge range of circumstances. Rules, quite simply, require less information.

The second reason is consistency. If bird behavior, say, were represented in the organism as a long list of what the bird should do for each particular set of circumstances, then genetic variation could easily change one item in that list in a manner that is inconsistent with the others. A rogue mutation might cause birds to flock except when they encounter a river, say. The flocking of migrating birds would then be hopelessly disrupted. In order for bird behavior to maintain its internal consistency as it evolves, it is much better to adapt by modifying the underlying rules.

The third is adaptability. Small changes in the rules can cause big changes in behavior. Therefore, behavior can evolve more rapidly, while remaining self-consistent, if it is stored as rules.

The fourth reason is that it's very hard to see how individuals could perceive and act on the overall characteristics of group behavior. Could bird genes contain instructions such as "the flock should cluster more tightly when a predator is sighted"? It's hard to see how. If this were the case, each bird would have to somehow be aware of the state of the entire flock; moreover, it would have to somehow become instantly aware if *another* bird has sighted a predator. It seems far more likely that the individuals obey such rules as "try to stay close to your immediate neighbors" and "if *you* see a predator, stay even closer," or even "if your immediate neighbors are clustering around you more densely than usual, start to watch for predators." Rules such as these, which do not require improbable acts of omnipotence from individual birds, but which can generate the observed collective behavior, make far more sense: local rules, rules that involve only what an individual can reasonably be expected to perceive.

Circumstantial evidence supports this contention. We know that human beings in crowds are generally unaware of the overall behavior of the crowd because crowds can sometimes become dangerous to the people in them *without anybody realizing until it's too late.* Most countries have a catalog of tragedies involving the unintelligent behavior of crowds. Yet the individuals in these crowds are intelligent. However, they are aware only of their immediate surroundings; they can't see around corners. Another likely example is what the British police perpetually refer to as "motorway madness" when large numbers of cars drive too fast in fog. I suspect it is a collective madness, not an individual one: Each individual driver is trapped in the crowd of cars and has to go along with it. Yes, there *may* be an unseen obstacle ahead, but there is definitely a large truck not far behind, going too fast. If you slow down, you know that the truck could easily crash into you. If you keep going—too fast for the weather conditions, but at the prevailing speed of the surrounding traffic—you are probably *safer.*

When we do try to understand group behavior in living organisms, the need for mathematics becomes overriding. Why? Because group behavior involves not just organisms, but also interactions among organisms: the behavior of *systems.* Mathematics teaches us that such behavior can often be wildly counterintuitive—and offers the prospect of improved intuition. So between rules and the resulting behaviors lies an expanse of intellectual territory that mathematicians are claiming for their own. We certainly do not understand the patterns of behavior in animal societies nearly as well as the animals themselves do. However, we *are* beginning to understand that much animal behavior has a mathematical basis; that a great deal of it is considerably less miraculous than appearances might suggest; and that a set of mathematical rules, hard-wired or otherwise programmed into animals' nervous systems, can generate far more subtle behavior than we would anticipate.

The theme of this chapter, then, is *rules.* What kind of behavior do you get from obeying—slavishly and literally—a few rules? How much of the social behavior of living creatures can be explained in this manner?

Mark Tilden, a self-styled "robobiologist," makes robots. His laboratory at Los Alamos contains "Robot Jurassic Park"—about 200 tiny solar-powered robots. He has designed and built a solar-powered robot that follows sunbeams around a room. Once inside a patch of sunlight on the floor, the robot appears to go to sleep. As soon as the patch moves on, it appears to wake up, nose around for a few seconds, and follow the patch, going back to sleep as soon as it is safely back in the daylight. It's not a big robot, but in the art of sunbeam pursuit, it does a job that a cat would consider admirable. Observing its behavior, you would imagine that some pretty smart electronics and some very fancy programming had gone into making it track sunbeams: It must be able to recognize the boundaries of the sunbeam, calculate how to move to get well inside the boundaries, and track the boundaries as the sun moves across the sky and the shadows move with it. Not so. This is a remarkably stupid robot. It doesn't even know what sunlight is. All it does is constantly to scan its electrical input from the solar array, and follow three rules:

1. If your solar cells are not generating more than some threshold level of power, spin at random and move 10 cm (about 4 inches) forward.

2. If they have been generating power above threshold for less than five seconds, move forward in a straight line at constant speed.

3. If they have been generating power above threshold for more than five seconds, stop.

That's it.

Until the robot finds a sunbeam, it obeys rule 1, and it spins, moves, spins again, moves again, zigzagging randomly across the floor. Eventually, it encounters a sunbeam, at which point rule 2 kicks in, and it keeps going in the same direction. If it's still in sunlight after a few seconds, rule 3 takes over, and it goes to sleep. As soon as the sunbeam moves on, the random zigzagging begins anew—and the same happens if by some mischance the robot runs across the corner of its sunbeam back into the shade.

The robot looks as if it possesses an intelligent and highly adaptable ability to follow patches of sunlight. Actually, however, all it's doing is obeying three rules. There are single-celled creatures that respond to light, and they may well obey a similar set of rules. Other single-celled critters follow chemical gradients, heading into regions of higher concentration like a dog homing in on an especially exciting smell. Ditto.

The moral: Don't underestimate the efficacy of rules. My next example gets closer to real organisms: It is the brainchild of Fritz Vollrath,[1] a zoologist. It concerns not a collection of animals, but just one: a spider. Nevertheless, it again addresses the relationship between general rules of behavior, and the specific behaviors that appear when those rules are applied in some particular set of circumstances.

Most of us have been impressed, at some time or another in our lives, by the cleverness of a spider, spinning its delicate web, stretching lines of thin silk between branches of a bush or stalks of grass, weaving a net with which to ensnare unwary flies. Vollrath thinks he knows how they do it.

Together with a student, Thiemo Krink, he has created mathematical "cyberspiders" that build realistic webs by following a few rules. They are not actual robots; they are computer simulations. With the right engineering, however, you could make a robot spider and set it to work catching robot flies—assuming you could get tiny robots to fly, which is beyond the present state of the art. Vollrath deduced his system of cyberspider rules through a mixture of observations of real spiders and experiments with computers.

The versatility of spiders is astonishing, the variety of web designs more wonderful than the assortment in any mail-order catalog. There are custom-built webs, off-the-peg webs, webs carefully tailored to their surroundings. There is the traditional web that looks a bit like the wires on a dartboard, spun by spiders such as *Nephila clavipes,* the orb weaver spider (Plate 18). There are many other designs of web, too (Figure 81). Some are so simple that they look somewhat pathetic: *Magrammopes,* the tropical stick spider, goes fishing for flies with a single line. *Hyptiotes,* a devotee of minimalism, builds a triangle. On the other hand *Scoloderus,* the ladder-web spider, is an accomplished artisan that weaves a web like a ladder, as you've no doubt guessed from its name. *Theridion* builds a web like a cat's cradle. *Dinopis,* the ogre-faced spider, hangs head down holding a tiny web in its front legs, lunging at any fly who approaches too close, like a child on the seashore with a fishing net. *Mastophora,* the bolus spider, swings a single strand with a heavy blob at the end, like the gauchos of the Peruvian plains. *Latrodectus,* the infamous black widow, builds a web that is narrow at the top, broad at the bottom, and reaches down to the ground—a funeral drape for the unfortunate insects that blunder into it. *Agelena,* the funnel-web spider, builds a funnel-shaped web and waits for its prey to fall into the funnel. And *Liphistius,* the trapdoor spider, lines a deep hole with threads, covers it with a trapdoor, and extends sticky tripwires out into the nearby terrain, to serve as an early warning system. *Liphistius* is a living fossil: Its ancestors built similar webs 380 million years ago. It is believed that all modern spiders are descended from ones that built webs like *Liphistius*'s.

Spiders are tireless workers: Most spiders construct a new web every day. They usually have to, because struggling flies can cause a lot of

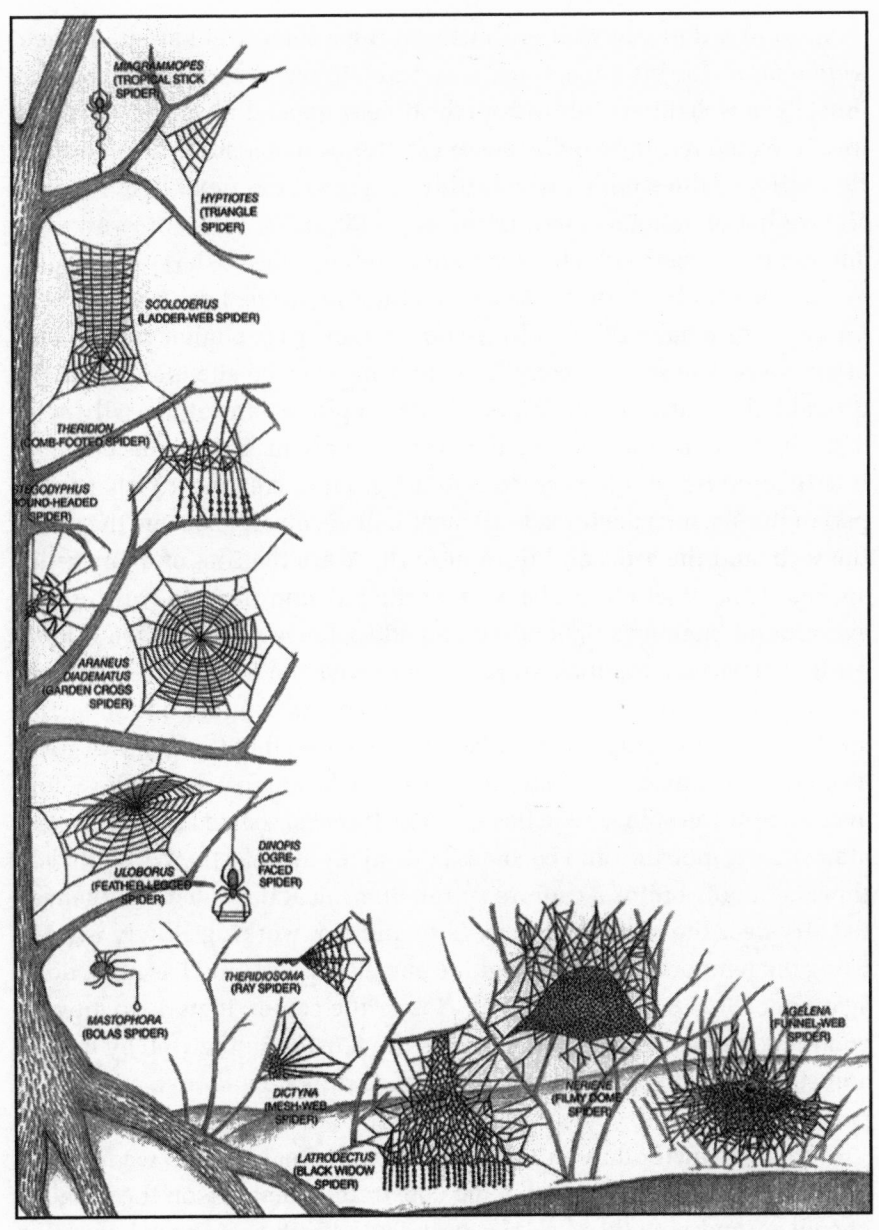

Figure 81
A variety of spiderweb designs.

damage to a delicate web. The common garden cross spider *Araneus diadematus,* for instance, builds around 200 webs during its short lifetime. Each web differs subtly from the others, depending on the surroundings in which it is to be built. These variations demonstrate the enormous flexibility of the spider's web-building system: too flexible to be just a simple list of actions coded in the spider's genes. Rules, however, are another matter entirely. The shapes of webs are clues to the spider's rules.

Here's what Vollrath thinks the creatures do. Here, I describe the rules for the traditional dartboard (*Nephila claupes*) web; similar rules govern other webs. The spider begins by exploring the web site, trailing its silk thread behind it. When it finds a suitable place—and it is easily satisfied—it strings a thread between two shoots or branches, climbs out along it to the middle, and lowers itself to the ground, creating a Y shape. The part of the Y where the threads all meet will eventually become the hub of the web, and the arms and upright of the Y are the first of many radial spokes. The spider climbs back up to the hub and starts to walk around and around, making a tight spiral and adding new spokes. Then it forms a widely spaced temporary spiral, which plays the role of scaffolding for the final web and is torn down as the web nears completion. So far, the spider has been using silk that is not sticky—so that it can move freely around its still incomplete structure. Nonstick webs catch few flies, however, so now the spider switches to sticky threads, coated silk that absorbs atmospheric moisture and becomes both sticky and elastic. With its sticky threads, it now builds a capture spiral. Starting at the outside of the web, usually near the bottom, it zigzags to and fro, working slowly inward, using the temporary spiral as a guide and as support, and tearing it down again once it is no longer needed. Nearer the center, it starts to move in complete circles. Finally it fine-tunes the tension in the web by making adjustments at the hub, settles down, and waits for a fly to come into its parlor.

That's the method in outline. At the next level of detail, we must ask how the spider plans its web, what constraints there are on the numbers of spokes, and how it spaces the strands; Vollrath has worked out effective rules for those, too. For instance, when it comes to deciding how many spokes to make, the spiders seem to be unhappy if adjacent spokes are separated by too large an angle, so they add new ones until all angles are acceptable. They measure the spacing between the threads of the capture spiral with their front two legs—just like a tailor measuring cloth.

Spider genes can easily equip spiders with rules of behavior—say by programming the architecture of the spider's nervous system. Which genes do what is a question for genetics, but the effect of those genes on behavior involves the link from rules to their consequences, and to understand this requires mathematics: as usual, a partnership. Vollrath's ideas show us what kinds of rules the genes must encode.

Rules encoded in genes can evolve—and again we need mathematics to understand the resulting patterns. Indeed, Vollrath uses a mathematical technique known as a "genetic algorithm"[2] to study how the spider's system of rules could evolve in the first place. The adjective *genetic* indicates that the mathematics borrows a useful trick from genetics, not that it works directly with spider genes.

An *algorithm* is a method for solving problems on a computer—a list of the precise steps required to carry out a desired computation, a list that comes with a guarantee that the computation will stop, with the correct answer. Most problems can be solved in many different ways, and the object of much computer programming is to find the most effective algorithm—the one that solves the problem the most quickly, using the least amount of memory, or satisfying some similar criterion. A typical task is to solve the "traveling salesman problem." The salesman has to visit every town on some list, but the order in which he visits them is up to him. What's the shortest route that visits every town? When the number of towns becomes even moderately large, it is impossible to solve this problem by listing all possible routes and seeing which is shortest: The number of routes becomes far too big. Some other strategy has to be used to solve the problem—or, at least, get close to a solution, because for practical purposes, it's more important to get a reasonably short route quickly than it is to spend aeons trying to get the absolutely shortest one. The genetic-algorithm approach starts with several randomly chosen routes. Then it chooses two of them at random and sees whether it is possible to combine the best features of both to get an improved route. Maybe one route is a lot shorter than the other when it comes to towns in Colorado, but a lot longer for towns in Utah. If so, a mixture of the two routes can be found that performs well in both states. In effect, this method cross-breeds the two routes and sees how good the offspring are. Shorter offspring are retained, longer ones are killed off—removed from consideration. This algorithmic approach is a deliberate analogue of Darwin's concept of natural selection, and it has the same effect. By repeating the process many times, extremely effective algorithms can be *evolved.*

In contrast, it is often difficult to *design* good algorithms from scratch if the range of alternatives is too big. It is in just these circumstances that genetic algorithms come into their own. They have many practical uses: For instance, they are often used by companies that trade on the money markets, where they allow currency dealers to evolve effective strategies for buying and selling, with the aim of making as big a profit as possible.

To see how rapidly a rule-based system for web building can evolve to build effective webs, Vollrath applied genetic algorithms to his cyberspiders. Working first with Nick Gotts and later with Peter Fuchs, he equipped the cyberspiders with their own "genes"—computer codes that represented their web-making rules. Then he cross-bred them, mixing bits of codes from two distinct parents, and imposed a form of natural selection: The ones that make more effective webs (effective at catching cyberflies, that is) survive to pass on their genes; the ones that make really bad webs do not. Using genetic algorithms, Vollrath found that it takes no more than 50 generations to breed highly effective cyberspiders. The money markets yield their secrets less readily, I gather.

The message here is mathematically exciting but biologically sobering. It is that *apparently* complicated and flexible animal behavior patterns can be generated by much simpler, and more rigid, rules. Evolutionary selection is based on the effectiveness of webs—but selection works by eliminating spiders whose web-building *rules* produce less effective webs.

What drives spider evolution is the rules, not the webs—and the same probably goes for much animal behavior, individual or collective. This message is exciting, but also sobering. The excitement arises because of the possibility of explaining complicated behavior in mathematical terms. Newton discovered that a particular kind of mathematics, calculus, could explain every intricate detail of the previously unruly movements of the planets. Perhaps there is a calculus of behavior, too. The sobering message is that nature may not be quite as marvelous as it seems to be—though it is surely marvelous enough, and debunking a few behavioral patterns makes only the tiniest of dents in nature's richly deserved reputation for subtlety.

There is also a message for biological research. If behavior really is represented in an organism by rules rather than by the effects of these rules—as seems likely—then knowing the organism's DNA is only one step toward understanding how the organism behaves. DNA may *determine* the rules, but the behavior that is generated by those rules is at best *implicit* in the DNA code. Making it explicit will require an understanding of that calculus of behavior—whatever it may be. Deducing behavior from rules is, almost by definition, a question for mathematics. Maybe the needed mathematics does not yet exist—but you'll never solve the problem of animal behavior just by sequencing DNA, looking for proteins, or otherwise focusing solely on molecular mechanisms.

Rule-based systems really come into their own when we want to understand animal colonies and societies. We've already encountered one dramatic instance in a relatively simple organism—the slime mold, which follows an elegant mathematical route to set up suitable conditions for its reproduction. I continue with a more complex organism, the ant, and eventually, we'll work our way up to what we humans—in our totally objective and dispassionate way, of course—view as the most complex creatures on Earth: ourselves.

Ants are intriguing because they are social creatures, forming huge, organized colonies. Insect colonies are often organized at a level that seems well beyond the capacity of such creatures—for example, those termites and their air conditioning. In 1987, Douglas Hofstadter[3] pointed out the "mysterious collective behavior of ant colonies." He remarked that they build huge and intricate nests, even though the roughly 100,000 neurons of an ant brain probably do not carry any information at all about the nest structure. How, then, does the nest come into being? Where does the nest-building information reside? Hofstadter thought that it must somehow be spread about the colony, in the caste distribution, the age distribution, and the individual ant body. I wonder whether "information" is the right concept here: Maybe most of what seems to be missing is taken care of by built-in rules of ant behavior, which utilize available patterns provided by mathematics.

In 1993, Brian Goodwin, Ricard Solé, and Octavio Miramontes[4] set about modeling the collective behavior of ant populations, using a computer programmed with mathematical rules for individual ants and for their interactions with each other. The most striking feature that emerged from their simulations was a coherent oscillation of activity within the nest. At first, the activity would be low—not many ants were moving about, and those that were moving were not moving very far or very fast. Then the activity would start to pick up, with more ants scuttling around the nest. The level of activity would increase even more, until the whole nest was a maze of scurrying ants. Then, after a time, the level of activity would decline, until again the nest was relatively quiet. This oscillation in the activity level was close to periodic: The graph of activity repeated much the same changes over and over again at approximately regular intervals. The whole phenomenon became more and more obvious as the number of ants increased (Figure 82).

This oscillatory behavior is very, very curious indeed because it is not at all clear where the oscillation comes from. For a start, there was no periodicity visible in the rules themselves, or in their immediate

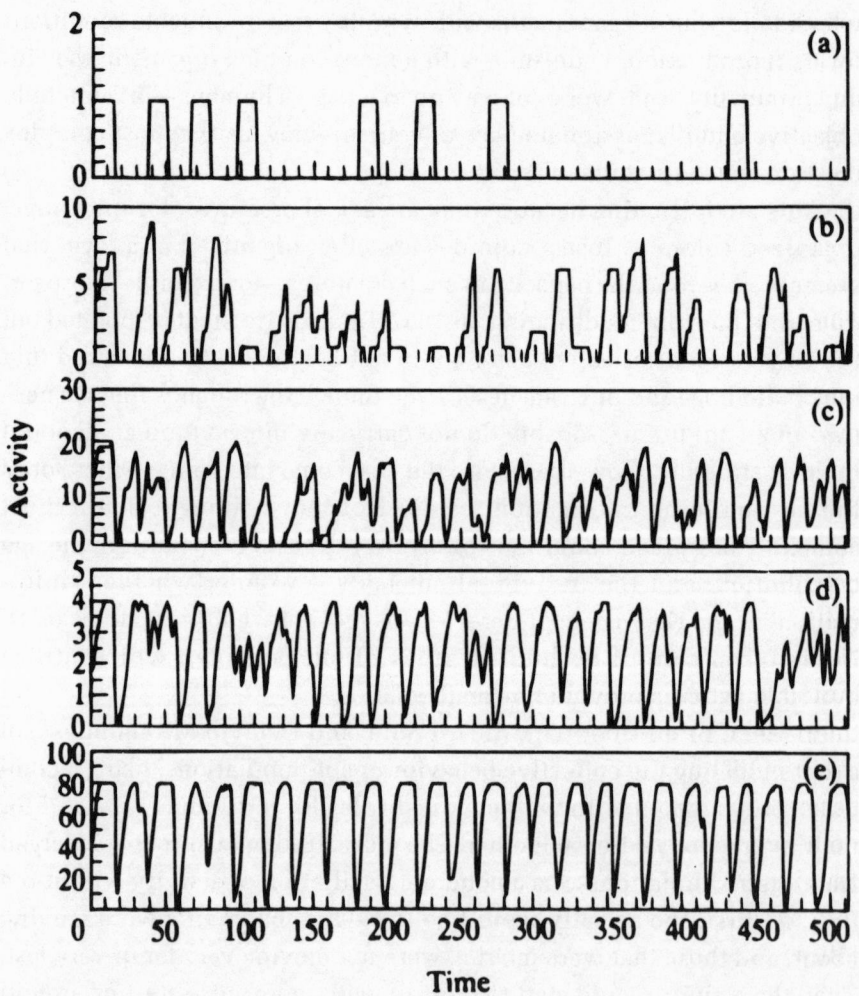

Figure 82
Spontaneous oscillations in model ant colonies.

consequences. Moreover, *no individual ant followed a periodic cycle of activity.* Compare this with what happens to traffic in any city on a working day. Before dawn, activity is low. Then a few cars appear on the streets, driving to work. Soon the rush of workers produces gridlock. After about nine o'clock, the activity begins to die down. It continues at a moderate level all day, picks up again in the late afternoon, dies down slowly, and pretty much ceases sometime after midnight. Then it all repeats the next day. At first sight, this pattern seems much like a description of the ants' nest, but city traffic differs from a nest in two crucial respects. First, the *rules* that underlie the behavior of the cars have their own daily periodicity: Everybody is driving to work, and they all work much the same hours every day. Second, the activity of each individual car is periodic: The same car drives to work at much the same time every day. So the periodic cycle of traffic activity in a city is built directly into the underlying rules and can be seen in each car separately.

The ants' nest is quite different. Nothing in the rules specifies an obvious time period; and no individual ant behaves in a periodic manner. The activity is *emergent* (not evident in the rules) and *collective* (a property of the nest as a whole, not a property of distinct individuals). Equally inexplicable oscillations of activity have been seen in experimental nests made by real ants. For example, N. R. Franks and S. Bryant observed large-scale oscillations in nests of *Leptothorax (L.) acervorum,* and B. J. Cole used a digitizing camera to observe *L. allardycei,* finding that the activity of the whole colony varied from low to high in a cycle that lasted between 15 and 37 minutes. Again, no individual ant behaved periodically, so here, too, the oscillating activity was collective and emergent. Not only was it not built into the DNA of individual ants; it wasn't even built into the *behavior* of individual ants. The oscillation is a property of the *rules* that genetics builds into individual ants. However, nothing in those rules tells any ant to follow a periodic cycle of activity—and because the rules govern the behavior of that ant, not of the whole colony, nothing in the rules can instruct the colony to follow a periodic cycle of activity, either. The collective pattern is again emergent: a consequence of the rules having a web of causality too complex for the human brain to follow in detail.

I think that the flocking behavior of birds, too, must be emergent. In 1987, the computer scientist Craig Reynolds[5] was wondering why birds formed flocks. Did they flock because that behavior was programmed into their DNA? Did they flock for evolutionary reasons (e. g., "it's safer")? Was flocking an emergent property of simple rules for bird behavior? Did birds flock for some other reason altogether? Reynolds devised a computer simulation in which virtual creatures known as "boids" flew through a virtual environment. Boids are instructed to balance conflicting urges: to avoid crashing into their neighbors, to stay close to their neighbors, and to move toward the center of the flock. Reynolds obtained highly realistic animations of boid flocking behavior, including such features as avoiding obstacles. Building on this work, Jessica Hodgins and David Brogan[6] found ways to reproduce the graceful motion of birds, fish, and other social animals that move in flocks, schools, or herds. They simulated a herd of robot pogo sticks (legged motion is often simpler to control if there is only one leg) and equipped them with rules for remaining together as a herd. Their pogo-stick herds successfully remained stationary, accelerated, turned, and negotiated obstacles, while "only a small number of collisions occurred." Hodgins and Brogan's main objective was to see how to concoct control algorithms for such herds, but they also demonstrated yet again that herding behavior is an emergent property, derived from rules but not explicitly built into those rules.

Earlier, I cited our own experience of moving in crowds—human herds—as evidence that collective behavior stems from individual rules, and that the individuals do not and cannot monitor the collective phenomena themselves. This point of view is creating a revolution in our ability not just to understand crowd flow, but also to predict it. The movement of crowds is an important practical problem for the designers of public buildings. In the April 1989 Hillsborough disaster in the United Kingdom, 96 people at a soccer match were crushed to death against barriers that had been intended to protect them. An effective technique for modeling crowd flow would make it possible to test alternative configurations of barriers, eliminate dangerous ones, anticipate possible problems, and work out ways of handling them safely.

Enter Legion, the brainchild of G. Keith Still.[7] On the Internet, Still is often known as Hari Seldon. In Isaac Asimov's science-fiction series *Foundation*,[8] Seldon was the genius behind "psychohistory"—a mathematical scheme for predicting the behavior of large groups of people. Legion does just that, in one particular case: It is a mathematical rule-based system that predicts crowd-flow patterns. The name is biblical— "My name is Legion, for we are many"—but Still took it from a more secular source, the cult British science-fiction TV comedy *Red Dwarf*.[9] Here the Bible was slightly misquoted: "My name is Legion, for *I am* many." Despite literary tradition, the *Red Dwarf* version is preferable: It makes the point that entire collections of creatures can behave as if they were a single entity.

The story of Legion began in 1992, at a rock concert. Still had been stuck in the same crowd for four hours, at the top of the stairs outside gate C of Wembley Stadium, still waiting to get inside. Wembley, Britain's premier sports venue, is best known for hosting the annual Football Association Cup Final, but this event was the Freddie Mercury memorial AIDS Awareness concert. There were 70,000 people, and the crowd wasn't flowing very quickly. Actually, Still realized, that wasn't true. It just wasn't flowing where he was standing. When he got home, he started to rethink, from the ground up, the problem of modeling a moving crowd. He realized that a crowd is a complex system, in which simple rules for individuals generate large-scale patterns. We may not be able to work out how this happens, but we can use a computer to implement the low-level rules, and a human mind can then look for any interesting large-scale regularities in the mass of data that results. That's how Still attacked—and eventually solved—the problem of crowd behavior.

At first, he tried to model crowds in virtual reality—but the available algorithms could handle only a few hundred people, not enough for a realistic crowd. He was up against a notorious problem in computer programming: intractability. When the virtual-reality system moved objects around, it had to do so without them interpenetrating, and it did this by checking each object against all the others whenever it was moved. So every time it updated the screen display, it looked at every possible pair of objects. As the number of objects grew, the number of pairs grew much faster. For example, with 10 objects there are 45 pairs, but with 100 objects there are nearly 5,000. Many problems of this kind are *NP-complete,* meaning that everybody believes them to be insoluble, even though nobody can prove it.[10] Fortunately, there was a way out: People do not think about everything in the same building before they move—they only react to what they see. And mostly what they are looking for is open space: They check the stuff that *isn't* there. So to move objects efficiently in virtual reality, they shouldn't interrogate *each other;* they should interrogate the nearby empty spaces.

The other big stumbling block was an almost total lack of understanding of the rules whereby individuals, moving in a crowd, make decisions about where to go next. Still had to work his way backward from crowd flow and then find the underlying rules. Staring at black-and-white video footage from closed-circuit TV cameras, he found himself doing time-lapse photography in his head. It was as if each person left a kind of wake behind them, a trail showing where they had just been. The result was a kind of order: patterns that broke up and re-formed, broke up and re-formed. . . . Sometimes two groups of people, heading in opposite directions, would meet head on. After a few seconds of apparently random jiggling, they would resolve into interpenetrating parallel lines, the way your fingers pass through each other when you lock your hands together. Alternatively, groups would crystallize, and lock solid, as they jammed up against a barrier, only to evaporate around the edges as the people found ways to free themselves from the obstacle. Mathematically, it was very different from any traditional flow problem, be it that of a fluid or of units through a network. It had a curious mixture of local and global features; moreover, some components, such as the people, were discrete units, whereas others, such as the directions in which they were heading, were continuous. It was a hybrid, and it could not rely on any useful existing body of known theory.

One fact, however, stood out: The *same* patterns formed in the same places, no matter what the event. The key to crowd dynamics is not the intricacies of human psychology, but the universal mathematical patterns formed by individuals that move and interact with each other in some surrounding geometry.

The geometry, in fact, was the most important feature. In March 1994, Still learned about *symplectic geometry,* an abstract but very important concept: It is the natural geometry of *motion.* He entered some sample formulas from symplectic geometry into his computer, trying to model the motion of individuals in crowds. He didn't see realistic crowds, but he did see something quite remarkable in its own right. Suddenly, on the screen, in vivid color, growing before his eyes, were the most striking, mysterious, and downright beautiful images that he'd ever seen (Plate 19). They were fractals, geometric shapes with structure on all scales, and one of them resembled an orchid—so he called them "orchid fractals." Orchid fractals didn't solve the problem of real crowds, as the fractals' rules were too rigid. Nonetheless, they proved that patterns oddly reminiscent of those in crowds can emerge from simple rules. In particular, orchids showed him how to program the geometry of crowds efficiently. Orchids were a form of self-ordering system—like boids, self-organizing into flocks, sweeping majestically around virtual obstacles in their self-contained computer world, breaking and re-forming. . . .

However, there was something about the analogies among orchids, crowds, and boids that didn't quite fit. The rules were different—not just in detail, but in spirit. To be specific, the rules for boid movement included, in mathematical guise, the explicit instruction "form a flock." That wasn't what *real* birds did. They didn't know they were *supposed* to form a flock—they just did it anyway. Their flocking behavior emerged from interactions among individuals; it wasn't built in as a common objective. The same had to be true for the patterns of movement in crowds. *The crowd doesn't know about its own overall motion.*

In a crowd, all individuals have some kind of objective—such as "get as near as possible to the stage" or "go to the bar." They have form: size and shape. They have internal variables, such as maximum and minimum speeds—senior citizens do not move as rapidly as teenagers. They obey rules of interaction—stay out of everybody else's personal space, try to take a step in roughly the right direction. Finally, and crucially, they function within an environment—the geometry of the surrounding building. The dynamic patterns of crowd movement follow from this structure. Within a few weeks, Still's ideas had jelled into a software package— Legion. Legion could handle crowds with a quarter of a million individuals, in real time. The people and their surroundings were stored in computer memory. Each person would interrogate its surrounding information space, decide whether that space was another person, a wall, or just empty space, and react accordingly. The precise rules of movement, which remain a commercial secret, are based on thousands of hours of observations of real crowds. Legion confirms that although the motion of any individual in a real crowd is irregular and unpredictable, the flow pattern is remarkably insensitive to changes in individual behavior. Both intelligent strategies to find the best moves and random strategies produce virtually identical flow patterns, even though individuals are moving according to very different rules in the two cases. The main determining factor is the geometry of the building.

Architects already walk through virtual-reality models of new buildings that exist only inside their computer. Now they can be accompanied on their travels by a veritable Legion of virtual companions, and they will be able to experience exactly what would happen if their brainchild caught fire, or if a barrier collapsed under the weight of numbers, as crowds of teeny-boppers struggled to greet their latest idol at the airport.

Systems such as Legion are beginning to open up a new scientific frontier—the emergent collective behavior of large groups of organisms.

Variants on Legion can model herds of animals, flocks of birds, schools of fish. They are ideal for modeling the intricacies of ecosystems, the overall collective behavior of which must again be a consequence of the rule systems of the numerous individual organisms. Legion makes real one of Hari Seldon's dreams from *Prelude to Foundation:* "a tool that might make it possible to identify what was good and what was bad for humanity. With it, decisions we would make would be less blind."

11

Reef Wars

A fungus is growing on an oaktree—it sheds more spores in a night than the tree drops acorns in a hundred years. A certain bacillus grows up and multiplies by two in two hours' time; its descendants, did they all survive, would number four thousand in a day, as a man's might in three hundred years. A codfish lays a million eggs and more—all in order that *one pair* may survive to take their parents' places in the world.

D'Arcy Thompson, *On Growth and Form*, Chapter III

Beneath the Pacific Ocean lies a hot spot where Earth's molten interior comes unusually close to the surface. Above it, volcanic islands climb from the ocean floor to the waves above. As the Pacific continental plate drifts, the hot spot occasionally breaks through in new places: Roughly every 200,000 years, it gives birth to a new island. The 8 islands of Hawaii are merely its most recent additions to a chain of 130 that stretches for more than 2,000 kilometers (about 1,200 miles) from the Kure Island atoll to the Midway Islands. The islands of the chain are born in a holocaust of fire, ash, and lava: the geological twinkling of an eye. Their death is more leisurely: Wind, waves, and rain eat them away, eroding them back to sea level—and below. The newest islands in the chain are the highest; many of the older ones are now completely submerged.

The second newest Hawaiian island is Maui—really the union of the second newest and a close predecessor, for Maui is formed from *two* volcanoes, Puu Kuki and Haleakala. Once separate, they are now linked by a narrow spit of eroded rubble.

Down the slopes of Haleakala, a series of minivolcanoes, side vents, and volcanic safety valves run out into the sea. Three kilometers (almost 2 miles) offshore lies the submerged crater of Molokini, a crescent moon of lava rising little more than the height of a building above the sparkling waters of the Kealaikahiki Channel. In nearby Maalea Bay, the humpback whales come to breed and frolic.

Above sea level, Molokini is little more than lava, grass, and scrub: a haven for seabirds and other small wildlife, forbidden to the casual visitor. Below the waves, however, Molokini is a riot of color, providing ground for corals, weird and convoluted, fanlike and fragile. Moray eels and octopuses lurk in the coral crannies. Spiky sea urchins bedeck the underwater landscape, grinding away at it with their voracious little mouths. Brightly colored fish flit among the rocks—parrot fish in rainbow hues, striped angelfish, spotted fish, vivid yellow fish like flattened lemons . . . stately patrols of deep blue fish, the police officers of the reef. Long, thin, gray trumpet fish gulp down short, fat, brightly colored fish—and then bulge, for days, like a python that has swallowed a rat.

A coral reef is a living, growing, dying thing. Primarily, it is a battleground. Different species of coral engage in hand-to-hand combat across a no-man's-land no wider than a human hand, a territorial war that has raged for millions of years and takes no hostages. More accurately, it is stomach-to-stomach combat, for corals have evolved the revolting trick of turning their digestive systems inside out and firing them at the enemy like a harpoon. They have two other weapons, too: chemical warfare and invasion.

Welcome to the Reef Wars.

Coral reefs—not this one, for it is protected, but most of the others dotted around the world, mainly in the Pacific and Indian Oceans—are engaged in another kind of warfare: one with people. People attack them with pollution from land and ships, deliberate destruction to create marinas for tourist boats, and incidental destruction by desperate people who dynamite the reefs to stun fish and other organisms. The dynamiters then sell their prey to collectors with marine tanks, receiving only rock-bottom prices for the havoc they wreak, while retailers reap the profits.

A coral reef is not a simple thing. It is a complex ecosystem, a web of interacting organisms. *Ecology* is the branch of biology that studies how many different organisms interact with each other in the natural environment; it is a relative newcomer to the pantheon of sciences. The same word is often used to describe the system formed by living organisms in their environment, but it is less confusing if we refer to the latter not as an ecology, but as an *ecosystem.* The reef ecosystem works: For hundreds of millions of years, coral has successfully been building reefs. Therefore, it must be a system that can, to some extent, withstand the ravages of chance destruction. Ecological webs are robust, not fragile; they can often survive the temporary loss of some of their strands. Can they survive wholesale onslaught, though?

In today's world, ecosystems that are unduly sensitive to human intervention will no longer survive on their own. Ironically, the only plausible way to keep them alive is though more human intervention. The way to save ecosystems from destruction is to manage them.

Why not just leave them alone? It would be wonderful to adopt this "deep green" solution, to ban all human activity, turn back the clock, and let nature reassert itself. Nature's clock will not turn back, however—certainly not easily. A hillside that has been deforested by logging and has lost its topsoil will not readily regrow its forest. In any case, people do not take kindly to being forbidden access to large, ecologically rich regions of a crowded planet. Try to stop a desperate parent in a war zone from dynamiting the nearby reef to get money to feed his (or her) starving family. Explosives are easy to find—food in a war zone is almost nonexistent. Whether we approve or not, the best we can do to preserve our natural heritage is to manage it, to try to compensate for past and present human interference, to try to undo the harm that we wreak.

There, we run into a serious problem. The road to hell is paved, as they say, with good intentions. It is terribly easy to try to help an ecosystem survive—and end up by inadvertently destroying it. There are innumerable examples where organisms introduced for apparently benign purposes have ruined the native ecology: rabbits and cane-toads in Australia; gorse bushes, deer, and possums in New Zealand; aggressive African bees in South America, which interbred with native species to create the killer bees now invading the southern United States. . . .

In order to manage an ecosystem, we have to understand it—not just what it *is* doing, but what it *would* do if we interfered with it. In short, we need a way to map out the geography of ecological phase space.

Ecosystems are nonlinear; they often run counter to naïve linear intuition. For instance, allegedly safe levels of fishing can devastate spawning grounds and take fish species to—or beyond—the brink of extinction. Therefore, we must forge new tools that will allow us to come to a radical new understanding. It is not enough to bleat piously about the sacred web of life: We need to know how it works and what it will do if we disturb it.

Those tools are not yet in place. We have no laws of ecology as precise as Newton's law of motion—and I doubt we ever will because that's not *quite* the way to go. The important patterns of ecosystems are qualitative—computing them to 10 decimal places isn't what we should be aiming for. We are inventing new tools and refashioning old ones, and slowly we are coming to grips with the realities of ecosystems. Slowly, the deep patterns of ecological dynamics are beginning to come to light.

Some of these tools are classical—statistics, say, or differential equations. Others are new. One of the most successful is a cross between mathematics and computer games, known as a "cellular automaton." In such a hybrid, instead of teenage boys zapping aliens and racking up huge scores, we now have one coral species zapping another, or sea urchins zapping both—and the scores in the game reveal the underlying dynamic patterns. With that understanding, we can advise governments how to manage their sensitive ecosystems.

For example, in January 1991, during the Gulf War, Iraqi troops deliberately released large quantities of oil into the Persian Gulf. This damaged the ecosystem—but how seriously? Key players in the gulf ecology are kelp, sea urchins, and lobsters. Lobsters inhabit the kelp forests and eat the urchins; the urchins graze the kelp. If there are too many urchins, then they reduce the kelp to bare rock; lobsters and urchins die if they have not eaten within a fairly limited period. Every so often, a marine virus attacks the urchins, killing the lot, but new ones can reinvade the area from outside. In order to find out how serious the damage is, we have to unravel this web of interaction. As we'll see shortly, the new mathematical tools provide a simple and direct method for modeling this simplified gulf ecosystem and pinpointing the dangers of interfering—or not. Because the full gulf ecology involves many other organisms, of course, the results of the simplified model must be treated with caution, but with more effort, the same methods could handle much more sophisticated models. Despite being simple, this particular model raises a very interesting possibility for monitoring the gulf ecosystem by satellite—but more of that later.

Cellular automata were invented in 1953, when the great mathematician John von Neumann set out to discover the secret of life—well, at least, one secret of life: its ability to reproduce. He wanted to show that this ability is not some mysterious, ineffable aspect of some ethereal life principle, but a relatively straightforward and *universal* feature of ordinary matter. This matter, however, is organized in a very specific way. He hoped that his ideas might lead to the discovery of the mechanisms that earthly life-forms use to reproduce—this was just prior to Crick and Watson—and in a way, they did. Unfortunately, von Neumann published his ideas just too late for them to be useful to biologists: Crick and Watson got there first.

Crick and Watson worked out what DNA looks like and deduced how it might replicate. Von Neumann worked the other way, stripping the *general* problem of reproduction down to its essentials by asking whether it is possible, in principle, to build a self-replicating machine. He wanted to create a general mathematical theory of replication. His answer[1] showed that machines that can replicate really are possible. It also led him to the same abstract scheme that living creatures—on this planet, at least—actually use: to contain within themselves a coded representation of the process that gives rise to them. The work of Crick and Watson led to the discovery that this is how DNA works; it's von Neumann's mechanism, manifested in molecules. As I said, however, Crick and Watson, and those who took up their work and developed it, didn't need von Neumann's beautiful insight to reach that conclusion.

In order for his question to be interesting, von Neumann had to distinguish between trivial and genuine replication. Every replicative system operates in some environment, and if that environment is too rich, then the system has easy ways to cheat. For example, a letter is a self-replicating machine in an environment of people with photocopiers. To get around this problem, we must find some system that replicates itself stably and accurately while employing only low-level materials from its environment: for instance, a robot that can dig up ores, smelt them into metal, dig sand and grow silicon crystals, etch its own chips, manipulate circuits and components, and eventually build another robot just like itself. Such a "von Neumann machine" would be a pretty complicated object—unless it was designed a lot more cleverly than my description suggests. One potential route is *nanotechnology,* with microminiaturized machines in a microscopic environment. Life went several levels of scale smaller, and it used molecules. Von Neumann headed into the abstract territory of the mind and came up with a mathematical scheme that replicated. In so doing, he invented a new mathematical concept.

Von Neumann's machine has two main parts. One is a manufacturing unit, which, given suitable instructions and enough raw materials, can build just about anything. The other is the list of instructions needed to build *itself*. At first sight, these requirements are paradoxical, for the following reason: If the instructions describe the complete von Neumann machine, then somewhere inside that machine, we must be able to find its instructions, so inside the machine's instructions, there has to be a description of those instructions, which means that inside that description, there is a description of the description of the instructions . . . and so on in an apparent infinite regress. However, von Neumann realized that this redundancy is not necessary. Why should the instructions contain a description of themselves, when you've already got the full instructions anyway? So he added a third component, an instruction copier. It didn't understand what was written on the instructions, and it didn't interpret or obey them. It just copied them.

Now we start with a machine consisting of a builder, an instruction list, and a copier. When the builder is given the instruction list, it makes a new builder and a new copier, and it bolts them together. Then it puts the instruction list into the copier and makes a new, identical instruction list. Finally, still obeying instructions, it places the instruction list in its proper place inside the new machine.

The trick for avoiding the infinite regress is simple but subtle: It is to treat the information in the instruction list in two different ways at two stages of the replicative process. First, the information is *interpreted* and obeyed. At this stage, what is written on the list is crucial: Mistakes don't just get copied, they make the builder go wrong. Then the information is *copied*. At this stage, the meaning of the list is irrelevant.[2] Mistakes do not affect the copying process itself—though they do lie dormant, ready to wreak havoc when the new machine starts to build its own self-copy. The genome has exactly this feature: During the development of an organism, it is interpreted; during cellular replication, it is simply copied.

Von Neumann could have stopped when he came up with this scheme, but there was still a residual question, "Can a mathematical system *really* do all that?" At the suggestion of another mathematician, Stanislaw Ulam, he came up with the aforementioned idea of a cellular automaton. The simplest type of cellular automaton consists of a line of squares, confusingly known as "cells." Two-dimensional arrays of cells, like vast chessboards, are also possible; in fact, von Neumann's was two-dimensional. You can even employ three dimensions if you want. Whatever the geometry of the cells, each cell is permitted a limited number of states, which in practice are conveniently represented by colors. At each tick of a notional clock, the state of a given cell changes, according to various rules that relate it to its neighbors. For example, one such rule might be "a pink cell situated between a yellow cell and a blue cell must turn green." Although such systems may appear very simple, it turns out that a cellular automaton can do anything that a computer can. The difference lies not in the capabilities, but in how much time it takes to carry them out.

Von Neumann invented a cellular automaton on a two-dimensional grid, with 29 states, and he laid down the grid in an initial pattern of around 200,000 cells. One region of those cells is a builder-cum-copier, the rest is an instruction list. The builder-copier region, following the rules of the automaton, extends a long tentacle out into empty space (Figure 83). Then it builds a copy of itself, following the instructions. Then it copies the instructions, adds them to the new region, and withdraws its tentacle.

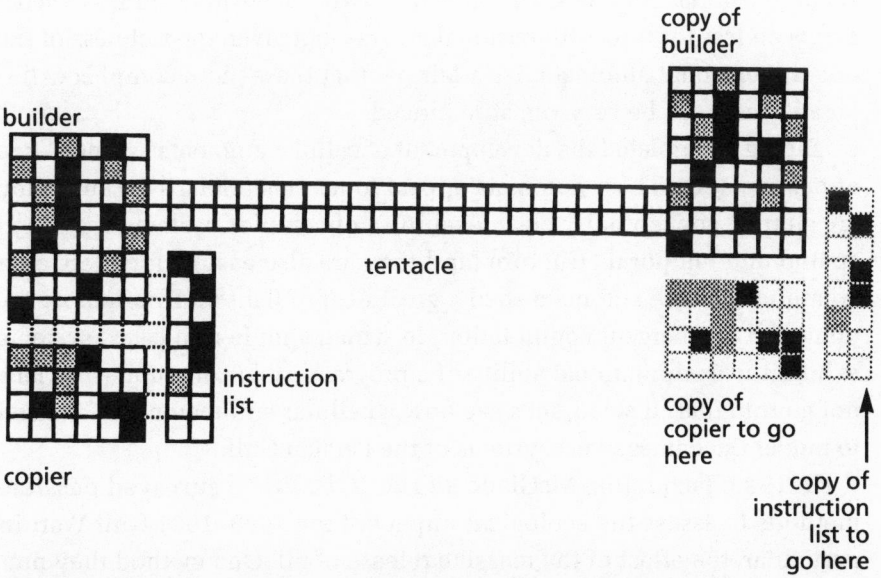

Figure 83
Von Neumann's self-replicating automaton extending its tentacle.

Just what *is* a cellular automaton capable of? In 1984, Stephen Wolfram (also famous for developing the software package *Mathematica*™) observed that cellular automata operating as a single *line* of cells produce four, and apparently only four, different types of behavior[3]:

◊ *Class 1:* Almost all initial configurations end up at a steady state in which all cells are in the same state forever, and that state is independent of the initial configuration (see, e. g., Figure 84a).

◊ *Class 2:* Almost all initial configurations end up at a steady state, or cycle periodically, but the state concerned depends on the initial configuration (see, e. g., Figure 84b).

◊ *Class 3:* Almost all initial configurations lead to a chaotic state, apparently unpredictable even though generated by the rules (see, e. g., Figure 84c).

◊ *Class 4:* Some initial configurations lead to complex localized structures; the automaton appears to be carrying out some computation (see, e. g., Figure 84d).

Class 4 automata are especially interesting: They can carry out the kind of computation required to implement von Neumann's scheme for replication. We know little about the general types of behavior that can occur in cellular automata that operate in two-dimensional arrays of cells, and even less for three-dimensional arrays—but given the richness of the one-dimensional automata, it's a fair bet that these more complex cellular automata can be very versatile indeed.

Biology stimulated the development of cellular automata as a new area of mathematics. These automata appeal to mathematicians because they are simple, conceptually clean examples of complex systems with both spatial and temporal structure (and they are also easy to simulate on a computer). These automata shed a great deal of light on Hopfield's phenomenon of emergent computation, in which simple rule-based systems exhibit the computational ability of a programmable computer—but let's not go into that. Instead, let's see how a cellular automaton can be used to model the damaged ecosystems of the Persian Gulf.

In 1993, Jacqueline McGlade and A. R. G. Price[4] surveyed possible methods to assess the ecological impact of the 1990–1991 Gulf War: in particular, the effect of the massive release of oil. One method they proposed was modeling by a cellular automaton. As an illustration of this method, they discussed a simplified ecosystem with three main types of organisms: kelp, sea urchins, and lobsters. They installed these organisms

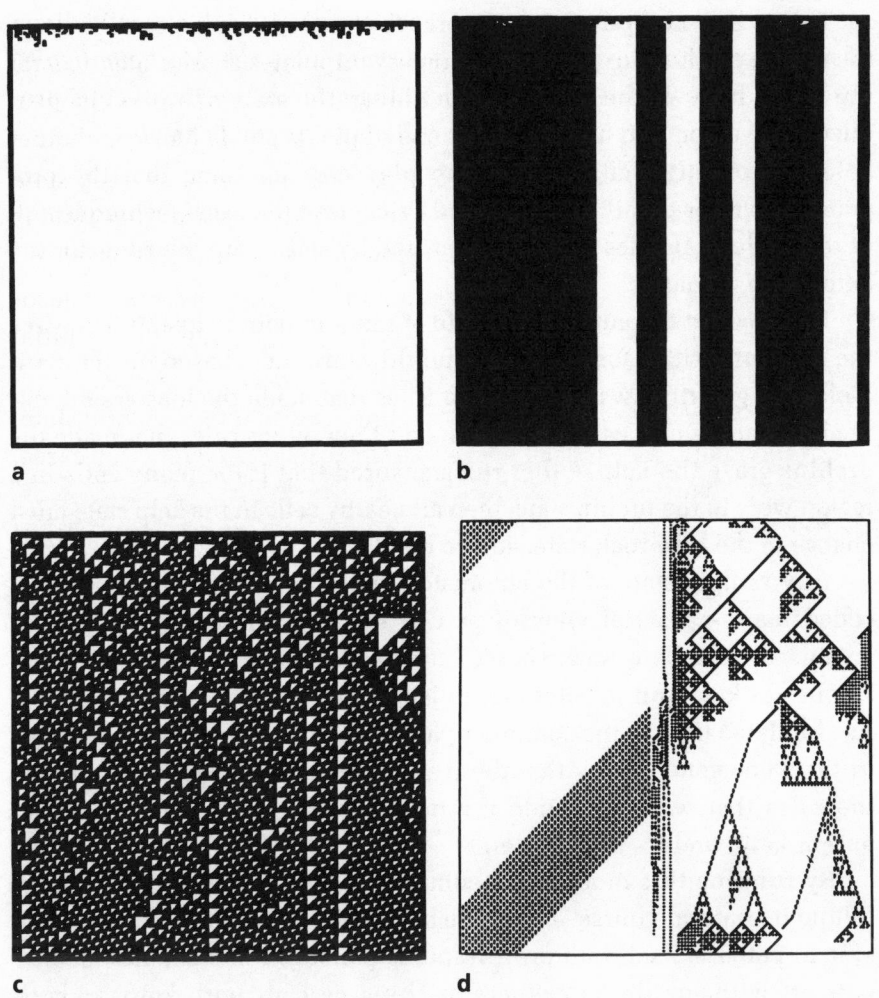

Figure 84
The four types of one-dimensional cellular automata.

in a custom-built two-dimensional cellular automaton. The grid of cells was a coarse model of part of the gulf geography—a coastline, a bay, some land, some sea. The possible states of a cell in the grid included a kelp state, an urchin state, and a lobster state. If a particular cell was in the lobster state, for instance, that just meant there was a lobster in the position defined by that cell.

This is how computer games work: To the programmer, each pixel on the screen has a state, and the program manipulates those states according to the rules of the game. To the player, the state of a pixel is visualized as a color, and associated groups of pixels create *Mario* and other beloved computer characters, who play out the game that the programmer's rules entail. McGlade and Price used the same technique: All of the ecological rules could be captured by setting up the rules for the automaton to match.

They seeded the automaton's grid at random with kelp and then gave the computer rules for how kelp should reproduce, based on standard biological growth laws. They set up rules that made the lobsters inhabit the kelp forests and eat the urchins; and they set up rules that made the urchins graze the kelp. Other rules ensured that if too many cells in a region were in the urchin state, then all nearby cells in the kelp state must change to the bare-rock state, and so on.

Any sample state of the automaton could be interpreted as a color-coded map—an aerial view of part of the gulf, in which the various organisms, and bare rock, showed up as squares of appropriate colors. Patches of kelp and lobsters were visible, bordered by beds of urchins. As simulated time in the computer passed, the model ecosystem changed in the same general way that the real one would—supposing it obeyed the rules that were built into the model, and ignoring any organisms except kelp, urchin, and lobster.

By running this model on a computer—letting the computer game follow its natural course—McGlade and Price showed that the gulf ecosystem patterns itself into prey–resource pairs. In some patches lobsters pair off with urchins; in others, urchins pair off with kelp. In such patches, the dynamics of the ecosystem simplifies into that of a two-organism system. As a result, the population dynamics become relatively simple. This patchy structure helps to prevent other species from invading the ecosystem, even under conditions for which it might seem that the prospects of successful invasion ought to be excellent. An especially significant finding is that the dynamics of all three species—kelp, urchins, and lobster—can be deduced from observations of just one species, say kelp. The reason is that the rules of the automaton link the kelp population—somehow—to the other two species, so by knowing one population, you can get a fix on the other two. You can't obtain answers obviously or easily, though: You have to run the automaton and observe its dynamics to find out just how the populations of the various organisms are related.

The important discovery is that you can monitor the ecosystem by observing the kelp population, using artificial satellite photographs. Even though lobsters and sea urchins can't be seen from a satellite, their numbers can be inferred from those of the kelp. So this particular mathematical insight leads to a novel, practical, but counterintuitive way to manage the gulf ecosystem. Don't worry about the lobsters and urchins— manage the kelp. Because the cellular automaton lets you deduce the entire dynamics of all three species from that of the kelp, you can use the mathematics to work out a sensible kelp-management strategy that will bring the whole gulf ecosystem back into its original state. You can also monitor the kelp remotely, with satellites, instead of sending expensive field teams to study the kelp in situ.

The Gulf War's depradations are but one example of how human stupidity and ignorance are endangering our planet. It is vitally important to the human race that we should learn to understand ecosystems. We grow forests for making paper, we take fish from the sea, we take tourist parties snorkeling over coral reefs, and we take fish from the reefs and sell them to people who own marine tanks. We plant wheat all over the midwestern United States. We drain swamps for agriculture, and we build dams to ensure water supplies. We do a million things of this kind, and every one of them disturbs a natural ecosystem. In addition, because we do not understand ecosystems at all well, we make mistakes. We fell trees on hillsides and then watch in horror as the topsoil slides off when it rains. Our wheat fields turn to dust bowls. We take so many fish out of the sea that their whole breeding cycle is disrupted, and suddenly there are no more fish.

Life is a highly interconnected system, but we know far less than we think about how those interconnections work. Our main hope is a mixture of field observations—to tell us how the organisms behave and interact— and mathematics, to derive the consequences of those interactions. Can mathematics really handle something as complicated as an ecosystem? If the answer is "no," then we're in trouble. I don't know of any miracle method for coming to grips with the complexity of an interacting system that is noticeably *better* than mathematics—though I can think of several that are worse. Zoology can teach us a lot about how an ecosystem behaves if we leave it alone, but far less about the consequences of changes. Genetics can track the flow of genes, but not the movement of animals. Economics can tell us how much our mistakes cost, but not how to avoid them. Computer simulation is a more attractive possibility—but that depends heavily on mathematics anyway: Where do you think the word *compute* came from? Either there is a general mathematical theory into which the simulation fits, or there isn't but there ought to be, and simulation is one step toward finding what it is. So to me, computer simulation counts either as mathematics or as embryonic mathematics; it is just one of the many techniques that people use to help them think mathematical thoughts. I am aware that this viewpoint is not held universally, but I believe it can be defended successfully.

Even if I'm right, this question must still be asked: Can mathematics tell us anything interesting about real ecosystems? It can, but, as always, you have to bear in mind the limitations under which all of science works. I would be astonished if it were possible to find a simple piece of mathematics that captured all of the intricacies of life on a coral reef, say—so in that sense, a coral reef is not now and never will be a slice of mathematics. Agreed—but that doesn't stop mathematics from being a useful way to understand a slice of life. As I've said before, I would also be astonished if it were possible to find a simple piece of mathematics that captured *all* of the intricacies of the motion of the bodies in the solar system. If you want to consider the *entire* solar system, then you have to catalog every body in it, and all the relevant characteristics of these bodies, such as position, velocity, shape, distribution of matter . . . which is too great a task for mere mortals. So the mathematics can never even get started on the real thing in all its glory. Despite these limitations, mathematics is an extremely important tool for studying the motion of the solar system. The trick is to use it on simplified models, not on the real thing. If the models are suitably chosen, then they capture key features of the real thing, and mathematics is what lets us extract the consequences of those key features.

It is—or should be—the same with ecosystems. The name of the game is to find simple models that incorporate the important features of some part of the global ecosystem and to use mathematics to draw whatever conclusions are implied by those features. However, if you dismiss all such conclusions on the grounds that you can find *something* in the ecosystem that doesn't match the mathematical model, then you're missing the point about how mathematics should be used. The aim is not perfection, but understanding—not exact descriptions, but insightful models.

One of the earliest aspects of ecosystems that attracted the attention of mathematical modelers was the growth—and decline—of plant and animal populations. The first half of D'Arcy Thompson's title says it all:

On Growth. He went into growth in a lot of detail. Today, when many species are in decline and some are close to extinction, this area of ecology has acquired a renewed significance. The concern is not just for exotic species: One of the big problems right now is the collapse of the cod population due to overfishing. One part of solving that problem is to understand how cod populations respond to changes in the rate at which they are caught; this at least tells us what targets for catches are viable. The science is only one aspect of the problem; politics also looms large. If we get the science wrong, however, then the politics can only make matters worse.

The earliest mathematical population model that I know about is the one that gave birth to the Fibonacci numbers. Recall that Leonardo of Pisa, commonly known as "Fibonacci" for dubious reasons, posed and solved a curious puzzle about the progeny of rabbits. There is no reason to believe that Leonardo intended it as anything other than a piece of interesting mathematics, but even so, his puzzle contains a serious message. Earlier, we solved that puzzle using Lindenmayer's grammar of treelike branching. An alternative method is algebra; with its aid, we can quickly discover that the solution is the Fibonacci sequence, and we can even use the rule that generates it (each number is the sum of the previous two) to find a *formula* for Fibonacci numbers. The formula turns out to involve the golden number,[5] which is not a great surprise because Fibonacci numbers and the golden number are known to be closely related. The upshot is that in Leonardo's model, the rabbit population increases, near enough, by a factor of 1.6 each season. If there are 100 rabbits this season, there will be about 160 next season, and so on. The number 1.6 is called the "growth rate." Because the growth rate is bigger than 1, the population increases; because the size at generation n is proportional to the growth rate raised to the nth power, the population increases—exponentially.

Exponential growth is very fast. If nothing happened to check the increase, then after 120 generations, the total volume of rabbits would exceed that of the known universe. This is the hidden message in Leonardo's little jeu d'esprit: In the absence of limitations imposed by resources or habitat, animal populations explode exponentially if the growth rate is greater than the breakeven point. By a similar analysis, the population declines exponentially if the growth rate is less than the breakeven point. *At* breakeven point, by which I mean that in the long run, one pair of adults in *this* generation produces one pair of adults in the next one, the population stays constant. In simple models such as Leonardo's, however, this balance point is highly unstable.

Of course populations always *are* limited by resources and habitat; but that's exactly the kind of quibble that leads nowhere useful if all you do with it is dismiss the model. The model predicts exponential growth before the effect of limitations takes hold, and this exponential growth explains many things. For example, it explains why bacterial infections can take hold very quickly, and why populations of pests—cockroaches, say—can come back very rapidly if you cease trying to control them, or if a resistant strain appears. It also makes it clear that any expanding population of animals is going to run into trouble when resource limitations start to bite.

The first semirealistic equation for population growth, known as the "Verhulst model," introduces an upper limit on growth in the simplest possible way. It assumes that unchecked population growth would lead to an exponential increase, but that this increase is mitigated by a scarcity of resources, in such a way that the population levels off neatly at the maximum sustainable value. This type of growth is called the "logistic curve," and it was one of D'Arcy Thompson's prize exhibits, occupying a large part of the first volume of his masterpiece. Even though logistic curves are classical, they remain of interest today, even at the frontiers of science. For example, they seem to be involved in mass extinctions.

Recall that mass extinctions are periods in which the diversity of organisms, as revealed by the fossil record, undergoes a sudden decline: Many species disappear. Subsequently, there tends to be a burst of renewed diversity as the development of other species is stimulated by the absence of those that have become extinct. The best known mass extinction is the demise of the dinosaurs, which occurred some 65 million years ago and was followed by the rise of the mammals. However, at least two other mass extinctions occurred in earlier times: one at the end of the Permian era, 240 million years ago; and another at the end of the Triassic, 200 million years ago. There are probably others, though their effect is less marked—and of course, the whole idea of mass extinctions is controversial because the fossil record is incomplete. In particular, it is not clear just how sudden or widespread the diebacks really were.

It would be nice to be able to say something more precise than "a lot of species died out at roughly the same time." In order to fit quantitative models to the observed numbers of species in a meaningful manner, it is important to build up good databases of ancient species—when they lived, when they became extinct. Until the mid-1990s, the standard databases were those compiled by J. John Sepkoski Jr. at the end of the 1970s. Sepkoski modeled changes in diversity—numbers of different species— by logistic curves, finding a reasonably good fit. However, in 1995, his conclusions were challenged by M. J. Benton, who compiled new data that (so he said) indicated not logistic but exponential growth. His data recorded not the number of species, but the number of families of marine organisms. In 1996, V. Courtillot and Y. Gaudemer[6] took a closer look at Benton's data and concluded that "diversification is determined not by exponential but by logistic behavior most of the time, and that this is interrupted by rare, brief . . . catastrophic events leading to mass extinctions." In fact, what the data show is that the nature of the growth depends on the time scale you use. Benton is correct that his data (Figure 85) exhibit roughly exponential growth over the entire period between

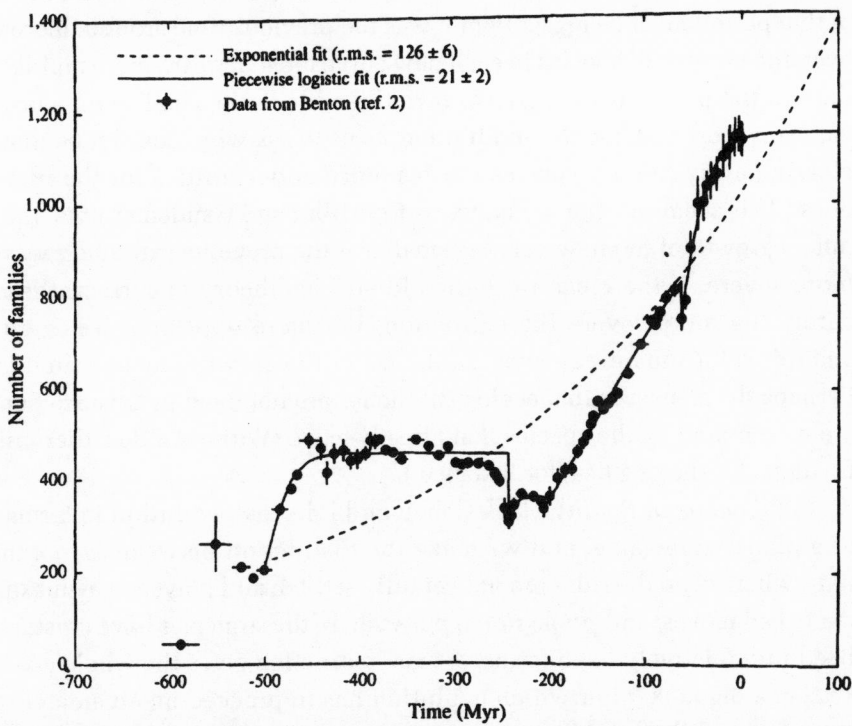

Figure 85
Diversity of marine species, with fitted exponential curve (dotted line).

500 million years ago and the present. On the other hand, the discrepancies between his data and a perfect exponential growth are quite large and highly systematic. If the curve formed by his data is chopped into pieces at the times of the three main mass extinctions, then the pattern becomes clearer: In each segment, growth is qualitatively like a logistic curve, but once the number of species has saturated, sooner or later there is a rapid decline—the mass extinction. After that, a new logistic curve begins. Courtillot and Gaudemer found the best quantitative fit for such a series of logistic curves, and they show—as is not unexpected from the diagram— that even taking into account the additional freedom offered by fitting several curves rather than just one, the multiple-logistic curve fits the data much better than a single exponential curve does.

Nevertheless, Benton is correct in saying that the overall changes in diversity are roughly exponential, and this assertion has implications for the separate logistic curves found by Courtillot and Gaudemer. It means that for each successive logistic curve, the number of species at the start of that period must be bigger than it was the previous time around; moreover, the number of species in each new curve must increase more rapidly than in the previous one, give or take the odd glitch. The fossil record confirms these statements, and it prompts us to ask why. Could it be that clearing away the old species creates more opportunities for the new ones? This seems a little too naïve; as Courtillot and Gaudemer note, the rate of growth of the new curve is smaller if the previous extinction was more severe. If the clear-away-the-old-species theory is correct, then surely the more severe the extinction, the *more* opportunities there should be for the new species, so the faster the growth rate should be. Perhaps the answer is that ecological niches are not fixed in advance but are determined by the species that already exist. (Without a dog, there is no niche for the dog flea, for instance.)

In *Figments of Reality*,[7] Jack Cohen and I discuss evolution in terms of a planetwide game, and we make the distinction between *amateur play,* where even the rules are not yet fully settled, and players may make many bad moves, and *professional play,* where the strategies have crystallized into relatively fixed forms. A mass extinction gives the whole ecosystem a big kick, after which evolution has to proceed on an amateur basis for a while, as nature experiments with the consequences of that kick. Because the niches for the new organisms depend partly on what survives of the old organisms, then—up to a point—the more species of the old organisms survive, the more niches the amateur newcomers can explore. (By the way, the old players that survive now return to amateur status because the rules of the game are all up for grabs again.) However, if too many of the old species survive, then you don't administer a big enough kick—so there is a breakeven point somewhere.

The logistic curve applies only to a single species—indeed, it lumps all age groups together, too, counting old animals along with young ones. More sophisticated population models employ many age groups and can handle the interactions of several species. Most such models are iterative, like Leonardo's: The population of a given species and age band in a given season is assumed to depend, by way of a specific mathematical formula, on the population values of all species and age bands during the previous season. An enormous variety of formulas have been proposed, each with its own pack of devotees, each attempting to capture particular features of the population dynamics. Those features include two that were well known to classical mathematicians:

◇ *Steady states,* where the population level is the same in each generation—a constant population

◇ *Periodic cycles,* where the population level changes in a repetitive manner—say from a high value to a low one and then back to the original high value, repeating the same two numbers indefinitely—a population that fluctuates predictably and regularly

However, there is a third possibility, one that only became known to science since the late 1960s, when nonlinear mathematics was applied to iterative equations:

◇ *Chaos,* in which the fluctuations of the population appear to be unpredictable and pretty much random

Actually, there are hidden patterns in chaotic fluctuations, but they are subtle enough not to make themselves readily apparent to a casual observer. Chaotic behavior has a degree of predictability, but only over the short term. For more about chaos, see *Does God Play Dice?*[8] The possibility of chaos changes our view about irregular changes in animal populations. Not so long ago, such changes would have been viewed as signs that some outside influence is at work—and our strong inclination would be to try to find out what. However, we now know that irregular fluctuations can be produced by the natural dynamics of the population itself; there may not *be* any outside influence behind the irregularities. By recognizing this potential for chaos, we can save ourselves a lot of wasted effort.

One of the big advantages of mathematical models is that they let you play about with different levels of detail. You can't take a real ecosystem and remove everything from it except kelp, urchins, and lobsters—so you can't tell whether those few species capture the important behavior. In contrast, you *can* take a mathematical model of an ecosystem and simplify it, step by step, to see whether its behavior changes in any essential way. In this manner, you can dig out the deeper laws that underlie ecosystem dynamics, removing unnecessary detail. Biology revels in details, mathematics prefers to get rid of them as far as possible; the two philosophies complement each other.

Mathematics lets us explore a whole range of possibilities and relate them in a systematic way. It even lets us introduce new factors into the equations, to see what difference they make. If they don't make a difference, then we can ignore them. If they do, then we've learned something useful—possibly something vital.

The first mathematical models of reef ecosystems revealed a puzzle. Leave aside human intervention, for a moment, and focus solely on the war between the different species of coral. If you model their battlefield, you find that in theory, one species—the one most suited to the environment of that particular reef—should win. Corals that manage to invade their neighbors' territory faster than their neighbors are invading *their* neighbors' territory will eventually dominate. In simulations, at high speed, the coral territories spin like pinwheels, chasing each others' tails—and the fastest spinner gobbles the lot.

Real reefs are quite unlike this scenario, however: They are incredibly diverse. Did the mathematicians get the equations for coral population dynamics wrong? No, that bit was fine. What *did* they get wrong, then? This is where the adaptability of mathematical models pays dividends. You can throw in any extra ingredients that you think might be needed— and see what happens. Maybe the diversity of a real reef is the result of diseases. As the reef heads toward a monoculture, the dominant species may become prone to epidemics. So we throw in some equations to represent the effects of disease organisms. We give it our best shot and build in what we know about real coral diseases. Do the results get rid of the incipient monoculture? No.

OK, so that's not it. Maybe it's parrot fish nibbling away at the coral. Add them in instead, again using good models for the population dynamics of parrot fish. Any better? Still no.

Hmmm . . . is it perhaps an environmental factor? Does the secret lie *outside* the reef altogether? A tropical ocean is not always placid. There are storms, cyclones, hurricanes—even tidal waves caused by underwater earthquakes. As far as the reef ecosystem is concerned, these strike at random. So we equip our mathematical model with random environmental disturbances, doing our best to match the statistics of the real reef's weather.

This one's *much* more interesting. Now the reef wars don't have a single winner. Before one coral species can mop up all the others, along comes a hurricane. It stirs up sand and silt and disturbs the reef ecosystem in dozens of other ways. Suddenly, the conditions that were making that one species superior no longer prevail. Then, before a new winner emerges, there's another disturbance to the reef, and so on.

It is the ever-changing environmental conditions of the reef that preserve its diversity. Paradoxically, damage by hurricanes is essential to maintaining the reef. Over millions of years, the reef has evolved a mix of organisms, which—among them—can survive whatever the outside world throws at them.

If we didn't know this and were thinking about how to manage a reef, it might occur to us that hurricanes are very destructive. We might build a huge seawall to keep them out. (I doubt anyone is contemplating this kind of thing because the expense would be vast, but you get the idea.) Our newfound understanding shows that such a step would be a disaster. It would destroy the rich diversity of life that makes a reef worth visiting.

In the management of forest ecosystems, we find a less hypothetical case in which too naïve an understanding of ecosystems led us seriously astray. After decades of trying to *prevent* fires from getting started in U.S. national parks, the prevailing wisdom has now swung full circle, and the current aim is deliberately to start small, limited fires. The forests *need* fires to remove tangled undergrowth and to regenerate saplings. If you don't have small fires, then the underbrush builds up and up, and one day, the whole forest burns down. Paved with good intentions, indeed. . . .

I'm not saying, "We shouldn't interfere with nature." We already *have* interfered, and we will have to do so again and again. After all, we live on this planet, too. What I'm saying is that if we interfere in a naïve way— or even choose *not* to interfere in a naïve way—then the counterintuitive mechanisms of ecosystems may very well kick us in the teeth. We must *understand* ecosystems, not just hold them in exaggerated respect—and to do that, we have to develop a whole battery of new, and often unorthodox, mathematical tools. Our investigative techniques have *got* to be mathematical; we don't dare risk experimenting on real ecosystems unless we already have a good idea of what's going to happen.

12

In Search of Secrets

Our own study of organic form, which we call by
Goethe's name of Morphology, is but a portion of that
wider still Science of Form which deals with the forms
assumed by matter under all aspects and conditions,
and, in a still wider sense, with forms which are
theoretically imaginable.

D'Arcy Thompson, *On Growth and Form,* Chapter XVII

Very well, then. What *is* life? Let me start by saying
what it's not. It's not DNA.

There was a time when the central concern of biology was organisms.
People studied zoology and botany, and they looked down microscopes.
They characterized life by its properties: reproduction, autonomy of
movement (i.e., freedom to *choose* what it does, or at least a strong semblance of such freedom), response to the environment, the ability (and
need) to find food, and so on.

Today the main area of biological activity has shifted to molecules.
People study biochemistry and molecular genetics, and they look at gels

for DNA sequences. They characterize life by its genome, by its DNA code sequence—by the fact that it *has* a DNA code sequence.

Like many scientists, I have very ambivalent feelings about this development. On the one hand, it is opening up the world of living creatures in a manner that is totally unprecedented. Scarcely a week goes by without some amazing new technique for exploiting the chemical mechanisms of life—DNA fingerprinting to capture criminals, test-tube babies, new ways to manufacture medicines, even Dolly the cloned sheep. Crick and Watson's discovery was truly epic: It has changed our view of biology completely, and it will change our world.

On the other hand, the emphasis on DNA and genes has been overdone. The images that have been planted in the public mind—DNA as the *Book of Life,* genes as the only feature of life that matters—are much too simpleminded. For example, we are subjected to persistent news reports, and announcements in scientific journals, to the effect that medical researchers have "located the gene for" some disease, disorder, or behavioral peculiarity—Alzheimer's, obesity, homosexuality, whatever. Nearly always, such reports are withdrawn a few months later, with the discovery that (a) the gene concerned does not always lead to the disease, and (b) there seem to be other genes involved, too.

The primary mistake here is to assume that features that make themselves felt at the level of entire organisms are caused by single genes. Occasionally, this is true: For instance, albinism is caused by a fault in a gene that makes a pigment. Such simple, direct chains of causality are the exception, however, not the rule. The secondary mistake is to assume that all diseases have genetic causes—when we know that many, such as chicken pox, do not. I shouldn't really say "all diseases," of course: Medical researchers are well aware of the cause of chicken pox. I should say "all diseases except the ones where we *know* that there is an external, nongenetic cause." This is just as bad, though: The unspoken assumption now is that if we don't *know* the cause, then it has to be genetic, and we should look for *the* gene that's responsible. Ignorance of an external cause, however, is not evidence in favor of a genetic cause. Not so long ago, we had no idea what causes chicken pox, so by the same token, *that* ought to have been genetic—but it isn't. About a year ago I was bemused by a biologist speaking on television about a hypothetical "gene for aging." The assumption was absolutely explicit: Aging is a disease and must therefore be caused by a gene. Wow! Find it, and immortality beckons! It didn't seem to occur to him that we may not need a gene to make us age: Aging could easily be a default option, a natural consequence of bodily systems breaking down. After all, a car doesn't need a gene to tell it to rust. Also, if there is a genetic cause for aging, isn't it at least as likely that a gene for keeping us young *switches off?*

And that's just the scientists. In the wider world, it is not at all unusual to be told—for example—that some people are having children "in order to pass their genes on to future generations." I suppose that a geneticist might be motivated in this manner, but I know that when I was deciding to have children, I didn't pay much attention to my genes at all. I blame this kind of nonsense on a widespread misunderstanding of the *selfish gene viewpoint,* which maintains that the only reason we exist is so that our genes can reproduce. The concept of the selfish gene is a clever and dramatic way to explain some peculiarities of DNA sequences, such as junk DNA, the function of which is unknown, but which still gets replicated along with the rest of the genome. However, it is equally possible to promote the *slavish gene theory,*[1] in which genes worry enormously about the survival of their organisms. (If the genes don't produce a viable organism, they die out, right?) Maybe the selfish gene has more going for it than I think, but even if it does, it is the story told *on the level of the gene.* The gene may in some meaningful sense focus on getting itself replicated—but that doesn't mean that humans focus on getting their *genes* replicated. Frankly, I don't greatly care whether my genes are passed on to future generations; my decision to have children is a social one, made for human reasons. I accept that at the back of that decision— deep down in our minds, where our consciousness cannot reach—are ancient imperatives that go back billions of years into evolutionary history. Evolution involves survival—and the ability to survive is transmitted by genes. Nonetheless, the idea that people actually care about their genes is ludicrous. They care about their families.

The same kind of dehumanization is creeping through many other areas of our society. Defendants on trial for crimes increasingly deny responsibility for their own actions on the grounds that their genes have caused them to commit murder, rape, theft, or assault. This is a difficult area because in extreme cases, there may be some truth to such claims. However, the idea that a predisposition to crime can stem from a single gene is largely nonsense. People commit crimes for many reasons, in combination; even if their genetic impulses tilt the balance slightly, normal human beings ought to retain enough control of their actions to override such tendencies. The genetic defense is in any case highly dangerous: If people are so slavish to their genes that they cannot prevent themselves from committing serious crimes, even if they try to, then they are far too dangerous to be permitted to walk freely in normal society. It's not their *fault*—but it wasn't their victims' faults, either. Further, it will be our fault, society's fault, if we buy the defense, "I am not an autonomous human being, I am a genetically controlled machine" but continue to treat the people concerned as if they *were* autonomous human beings.

I raise these issues not because I want to discuss ethical problems, but to show how deeply the image of genetic determinism has pervaded our society. That image, in turn, rests on the idea that genetics is *the* secret of life—the *only* secret of life. My belief is that it is merely the topmost layer of a series of secrets. You can sit DNA in a test tube forever, and it won't come to life. It won't even replicate. Life arises—on this planet at least—when the codes in DNA are plugged into a complex support network of physics and chemistry. The potential for life existed from the moment the universe began: DNA happens to be one route for making that potential real.

It is not, I am convinced, the only route. For it is not what DNA *is,* but what it *does,* that really counts. Let's take the phenomenon of life seriously, see its formalization as a mathematical challenge, and think about what kind of mathematics we need to develop an understanding of it. To start with, let's be fairly specific and think about life as it exists on Earth. Later, we can become more ambitious. An organism is made up from large numbers of cells, which interact with each other across their common boundaries, adjacent cell membranes. In this respect, the system is like a cellular automaton—but with two major differences from mathematicians' toy automata. First, the internal states of the cell are much more complex than black/white or green/red/blue. They resemble a complicated, multidimensional reaction-diffusion system. However, this system also incorporates elements of discrete computation, such as tubulin's ability to tear itself down and rebuild elsewhere. Second, the cells do not live on fixed grids. They are networked together, and they can move, grow, reproduce, or die. Moreover, there are complicated feedback systems, not just from the external dynamic to the internal one, but also the other way around. Each cell contains its own computer program—its genetics—which can react to what other cells in its neighborhood are doing, and which can constrain the cell's own response.

It would be possible to formalize this idea mathematically: The best attempt in this direction that I've yet seen is Agarwal's cell programming language (CPL). However, we have very little idea of just what general phenomena such a formalism implies. The best we can do right now is to choose rather simple systems along these lines, give them extremely simplified rules, run the computer, and see what we get. The results are fascinating, and the mix of flexibility and mathematical pattern is encouraging, but we can't spend the next 500 years inferring empirical principles from simulations. We've got to get to grips with such structures in a more conceptual way; we've got to do *real* mathematics with them.

The same kind of mathematical pattern extends both down into the microstructure and up into the macrostructure. The internal workings of a cell, say, have the same kind of intricacy, and a similar mix of ingredients, as those of a collection of cells. Now, however, instead of lots of cells interacting, we have the cell's own organelles doing the same. Even more clearly, when we move to the level of entire ecosystems, we have many different organisms interacting, in an even more flexible environment. The most astonishingly convoluted system of all is evolution: It has *all* of these ingredients, and they all feed back on each other, with the whole of the physical environment thrown in. The microstructure in one cell affects the organism in which it lives; that organism interacts with others in its environment, and also with weather, local chemical concentrations, currents in the ocean, impacting meteorites, whatever. On the other hand, the result of that interaction dictates whether that organism's own microstructure lives on by being replicated in other organisms. So the feedback runs in both micro-macro and macro-micro directions. Our mathematical understanding must come to grips with the unity of systems on different scales. Ideally, it must explain why that unity occurs at all.

I've generally downplayed genes as a source of explanation, but the role of genes is manifestly important, and we can't leave them out. Genes tinker with free-running mathematical systems and fine-tune their dynamics, so we have to build genes into our mathematical model, too. Now, it so happens that genetics is one of the best established mathematical areas in biology. Unfortunately, classical mathematical genetics is a pale shadow of the true evolutionary landscape. We need—and are beginning to get—*real* genetics, nonlinear models that adequately capture the true behavior of individual genes in the context of organisms, not just proportions of genes in a uniformly stirred gene pool. The days of linear mean-field theories are (or should be) long gone; nonlinearity and complexity are ascending.

In addition to genetics, we need an equally thorough theory of *phenotypics*—form and behavior in complete organisms. Then we need to rub the two theories up against each other to see what ignites. Also, if we want to understand the full implications of the results of rubbing those theories together, we have to face up to a deep, general problem, the problem of emergence. When you rub two versatile systems up against each other, you get what Jack Cohen and I call "complicity": Their effect in combination bears absolutely no resemblance to what they do separately.[2] *Complicity* is a kind of joint emergence: Again, the difficulty is not that there is no chain of causality, but that any such chain is far too complex to be grasped in its entirety, and there is no effective way to compress it. Once more, we run up against the need for a sound formal theory of emergent phenomena—the origins of what I much earlier called "God's book of patterns."

Let's look at some relevant patterns from the book. Crick and Watson were excited by one particular pattern in DNA: the possibility that it could be copied by a simple molecular mechanism. Copying is not restricted to DNA; it is a basic mathematical pattern in the universe. Von Neumann showed that an abstract mathematical system, a cellular automaton, can be persuaded to make copies of itself. The physical pattern here is relatively close to the surface. The laws of the physical universe permit the existence of systems that can copy other systems—universal replicators. When such systems are turned on themselves, they become *self*-replicating. This closure of the conceptual loop is crucial because self-replicating entities multiply exponentially. In contrast, when a universal copier replicates something else, the number of copies increases linearly. That is, if a photocopier can copy 1,000 documents per day, then in a year, it will make 365,000 copies. If, however, the photocopier was really a universal duplicating robot that could copy machines, not just documents, then it could copy itself. If it made just one copy of itself per day, then at the end of a year, there would be 2^{365} photocopiers—roughly 10^{110} of them: 1 followed by 110 zeros. Self-replicating devices *amplify* their powers because the powers are copied when the devices are.

Replication alone does not a life-form make—but it's a start. It is, after all, what got Crick and Watson excited about DNA's *double* helix. Another feature of DNA is that it can encode instructions. Again, this pattern arises from the physical universe. A code is a transformation, an input-output device that turns codes into whatever they mean. An instruction is a code sequence that results in some coherent event when so translated. DNA codes, in the first instance, are instructions for making proteins; proteins are the molecular building blocks for organisms. After this, our understanding of how organisms get constructed rapidly fades: We are pretty sure that special DNA sequences called "homeobox genes" help to organize which protein-building genes are switched on and which are not, but how the proteins are put into the right places is mostly mysterious. I don't mean that biologists aren't thinking about it, and I don't mean that they have no understanding at all—but the aspects of development that we understand are a pale shadow of what organisms actually do.

People talk an awful lot of nonsense about DNA as "information" for an organism—as if information has any meaning outside of a well-defined context. Ironically, this is a case where a *mathematical* image has been seriously oversimplified and misunderstood. The concept of information comes from communication theory, which deals with messages—sequences of code letters, say. The amount of information in a message is, basically, its length. Long messages convey more information than short ones. However, what's really important about a message is not the quantity of information, but its quality. "Two plus two make seventeen" is longer than "2 + 2 = 4," but it is also nonsense. More subtly, the concept of a message involves knowing that the appropriate constituents are its letters. Some device must be able to read those letters and *interpret* what they mean. A CD (compact disc) contains a lot of information, impressed as tiny bumps—but without a CD *player,* you'll never know what the music is. You won't even know that the bumps contain a message.

Bumps? What bumps? The *label* contains a message, in print. Sorry—bumps? By focusing on the DNA code, we are like someone trying to understand music who sequences the bumps on CDs, but doesn't take a look inside the CD player. My feeling is that the bumps on the CDs are a very simple trick, not especially interesting of themselves; all of the important action goes on inside the CD player. This is where the bumps are miraculously transformed into music.

Right now, the prevailing scientific wisdom is that the characterization of life in terms of DNA is considerably more profound than pre-DNA characterizations in terms of properties such as reproduction and response to environment. However, science tends to move in circles—except that when it returns to its previous position, it does so on a higher level, more like wandering up a spiral staircase. As attention starts to focus once more on what life does, rather than on what it is made of, the focus comes back to the properties of living systems. This time, however, we don't just want to catalog them; we want to understand them.

The test of true engineers is not that they can build one particular machine. It is that they understand the general principles behind machines and can apply them to whatever task they confront. Similarly, if we truly understood life on this planet, then we would be able to work out the generalities that underpin it, and—subject to practical constraints—we would be able to conceive of and even make new kinds of life: machine life, computer life, clay life, molecular life with exotic molecules, plasma-vortex life. By "life," I mean systems with behavior that is just as rich as that of organisms—systems having the same list of basic properties.

If someone, some*thing,* arrived tomorrow in a UFO with a robot that could do everything a human can do—not just mimic the appearance, but genuinely be autonomous, reproduce, and so on—I wouldn't waste my time arguing about whether the robot really was human. Such questions are a matter of definition, and they don't address the important issues. I'd be much more interested in finding out how the robot works, because that would offer new insights into how all such systems work—humans included.

In other words, there are two levels at which we can attack the question of life. The one favored by today's molecular biologists is to look at a specific realization of the general concept of life, see what it is made of, see what happens when you tinker with it, and make use of anything that emerges from such investigations. The one that interests me is different: It is to try to understand which properties of earthly life *generalize* to other realizations. What are the deep rules that life on Earth exploits? These rules include such processes as reproduction **and autonomous** behavior. DNA is just the trick that earthly life uses to exploit the deep rules—it's not the rules themselves.

Rules are the materials of mathematics, not of biology. Mathematics can certainly learn a lot from today's biology, but the point of view is different. Mathematicians always see any particular object as a special instance of something more general. Draw them a triangle, and they won't measure its angles to see which triangle it is; they will ask which of its features are typical of *all* triangles. Show them a frog, and they won't sequence its DNA; they will ask in what respects the frog is typical of more general, abstract systems possessing similar deep features. They seek to understand life in the abstract, not in the particular.

It is here that the true laws of life must reside. If we could understand them—and I'm not suggesting that we do, not yet, not for quite some time—then we would be able to characterize life in all of its potential forms. When we encountered aliens from the planet Zarathustra, whose biochemistry was based on ZNA (an imaginary long-chain molecule), rather than DNA, we wouldn't need to *revise* our conception of life to agree that the aliens satisfied it. Also, we would have a clear idea of what general features would be *common* to alien "zenetics" and human genetics, instead of getting hung up on the differences.

I still haven't told you what life is, but I hope you can see the kind of answer I'm after. Let's dig deeper. I mentioned the two celebrated features of DNA: replication and coding. It is striking that the really puzzling attribute of life—autonomy—does not seem to be visible in DNA at all. Physics and chemistry do not possess any obviously autonomous features. Indeed, on a philosophical level, it is difficult to see how a rule-based system can be autonomous. Its decisions must be fixed by the rules, so how can it *choose* what it does? Our current knowledge of genetics offers no understanding of autonomy at all.

I suspect that autonomy revolves around two distinct levels of description. One level is the level at which the rules reside. Here, the workings of the system are rigid and prescribed. The second level is on a much larger scale, that of emergent consequences of those rules. Because the consequences are emergent, there is no way for human beings to trace them back to the underlying rules. Such a connection exists, and the universe *produces* the emergent behavior as a consequence of obeying its own rules—but there's no shortcut that can enable humans to work out the actual chain of causality. As far as humans can see, the emergent behavior becomes disconnected from the rules. In particular, such features as determinacy, simplicity, and autonomy do not carry over from the rules to the emergent behavior. Also, of course, one of our human features is that everything we know about the external world is conveyed to us by our perceptions, so our limited perceptions are, in a sense, all that we can ever observe anyway.

The apparent autonomy of living organisms, then, is almost certainly an illusion, in the sense that the underlying rules are actually rigid. On the level of the organisms themselves, however, the workings of those rules produce emergent behavior with all the features of autonomy. The amoeba *looks* as if it is actively choosing whether to search for food. And so it is—but deep down, that choice is actually being made by a hidden list of rules. So complex is the link from rules to behavior, though, that even if you *knew* the rules, you would never be able to predict the behavior from them. In the same manner, we know the rules for weather. We also know, because of the phenomenon of chaos, that there is no way to predict the weather from those rules for more than four or five days into the future. The link is *there,* in the physical universe—but our minds cannot follow it.

If there is one single piece of new mathematics that could illuminate the general concept of life more than any other, it would be to make formal sense of what I've just said: to capture the abstract essence of an autonomous system, to deduce its properties, and to show that autonomy emerges from it. If we could do this, we could begin to develop a mathematical theory of the dynamics of autonomous systems—the general patterns whereby they change.

In the mid-1990s, Kauffman took one step in that direction. Another characteristic of autonomous systems—abstract life—is that they evolve. One of the most striking features of evolution is that every so often it leads to a complete change in the rules of the game. When oxygen built up in Earth's atmosphere, life evolved in a completely new direction, the eukaryotes. When flying creatures first appeared on Earth, utilizing the physical laws of aerodynamics and the fact that an atmosphere existed, they changed the game of life not just for themselves, but for all landbound life-forms, too. (For instance, the life of a field mouse would be very different indeed if there were no hawks.)

There is a basic law of statistical mechanics—the physics of gases, randomly bouncing collections of molecules—known as the "second law of thermodynamics." In its usual interpretation, it states that as time passes, systems without an external energy source will, with overwhelming probability, become ever more disordered. The interesting structure breaks down and vanishes. Autonomous systems, in contrast, seem to lurk in the near-zero-probability gaps where the second law fails to apply. Their typical tendency is to complicate their structure—spontaneously, inexorably, and (it seems) entirely naturally. Kauffman argues that autonomous systems become *more* complicated as time passes—and they do so as fast as they possibly can without falling apart. Their order *increases* over time, and it does so as rapidly as possible, subject to certain constraints.

This tentative "law of the complexification of autonomous systems" is probably untrue as stated, but it is a lot closer to the truth than is the second law of thermodynamics. The second law is a statistical law, asserting what happens with high probability, on the assumption that all possibilities are equally likely. The behavior of autonomous systems suggests that *not* all possiblities are equally likely. Such systems play merrily in the cracks of the second law—indeed they seek out the cracks because they function by exploiting those cracks. It's a bit like dangerous airports. You might expect that the accident rate would be highest at the most dangerous airports—such as Washington National, where each plane typically starts its takeoff along a runway while a previous plane remains on the same runway, after having landed. However, *the pilots know the airport is dangerous,* which makes them especially careful. So the statistics of accidents don't do what you'd expect. And neither does the statistical mechanics of autonomous systems. Maybe. It's an intriguing idea, anyway.

Even though biology has changed almost beyond recognition since 1917, D'Arcy Thompson's main message survives unchanged: Life is founded on mathematical patterns of the physical world. Genetics exploits and organizes those patterns, but physics makes them possible and constrains what they can be. Mathematical regularities are exploited by the organic world at every level of form, structure, pattern, behavior, interaction, and evolution. There is mathematics in the molecular scaffolding of DNA and in the long-term evolutionary dynamics of the entire global ecosystem, in the trotting of a horse and in the kelp grazing of sea urchins, in the glory of a peacock's tail, in the gaudy wings of butterflies and the markings on seashells, in the disposition of sunflower seeds, in the organization of ants' nests. . . . Moreover, there is evidence of mathematical features in numerous deep puzzles—the tubulin-building abilities of the centrosome, the development of an embryo, the division of a cell, the dynamics of an ecosystem—hinting that new kinds of mathematical insight may be just around the corner. Mathematics is not cast in tablets of stone: The interesting work, the fun, the intellectual challenge, and the practical payoff lie in chiseling new commandments onto new tablets, not in slavish obedience to the old ones.

What we have here is mostly a challenge, and only occasionally an answer. Whatever that challenge may be, it is not simply a question of writing down an equation of life and solving it. I doubt very much that any such thing exists. Finding exact solutions to ultimate laws is not a sensible role for mathematics. It's not math's role in physics, and it certainly should not be math's role in biology. The role of mathematics is to analyze the implications of models—not "nature red in truth and complexity," as Tennyson did not quite say, but nature stripped down to its essence. Mathematics pursues the necessary consequences of certain structural features. *If* a planet can be considered a uniform sphere, what would its gravitational attraction be like? *If* light behaves like a wave, what happens when two trains of waves cross? From this point of view, biological applications of mathematics should be treated in the same way. *If* animal movement, in simple situations, is governed by the natural oscillation patterns of a small network of neurons, what kind of network should it be? *If* the movement of cells in some circumstances is controlled by physical forces and does not greatly depend on complicated internal features such as mitochondria, what will the cells do? From this point of view, the role of mathematics is not to explain biology in detail, but to help us separate out which properties of life are consequences of the deep mathematical patterns of the inorganic universe, and which are the result of more or less arbitrary initial conditions programmed into lengthy sequences of DNA code.

Where do we go from here? Somewhere out there in the as-yet-unrealized space of future mathematical concepts, there are some extremely important ones about the nature of emergent form. In saying this, I am not trying to promote any particular school of current thought—chaos theory, complexity theory, catastrophe theory, cellular automata, neural nets, genetic algorithms, symmetry breaking, whatever. Those are merely hints, clues, partial and limited attempts. I think we have to go a lot farther.

It took 300 years between Isaac Newton realizing that differential equations can be used to understand the physical sciences, and the development of really adequate understanding of what differential equations are capable of. What is a differential equation? It is a mathematical description of what goes on *at a single point in space* as time passes. A single point.

Nearly everything interesting has spatial structure, though; each atom it occupies an entire region of space. That is what form is all about. Now there does exist a well-established extension of the theory of differential equations that applies to such problems: the theory of partial differential equations. Partial differential equations were invented to study fluids, heat, light, sound, electricity—and in those fields, they have been spectacularly successful. Every time you fly in a plane, you are exploiting the power of partial differential equations, for those are what engineers use to model the flow of air over a wing and to understand lift and stability in flight. Yet partial differential equations are much less well understood than ordinary differential equations. Virtually every interesting question about any specific partial differential equation is solved by numerical approximation on a big computer. In most cases, we have no idea why the computed solutions behave as they do; we merely have the computer's assurance that this behavior is a logical consequence of the initial equations. It works, it's practical—but I don't find it satisfactory *as a way of explaining nature.* It doesn't tell me *why* partial differential equations produce the kinds of behavior that apparently they do. It doesn't help me solve the next problem about pattern formation in a new partial differential equation: All I can do is put that on the computer again, solve it approximately, and be just as baffled as before when I am presented with the result. "The computer says so" may provide answers to *specific* questions, but it does not provide insight. This is the mathematician's true instinct: never to be satisfied unless results are derived in a manner that makes them appear inevitable. For instance, natural selection is a biological insight of the kind that mathematicians respect.

I am not trying to insult theorists of partial differential equations. They have performed miracles of understanding. It's a tough problem, though, and I'm asking for understanding at a very deep—perhaps unreasonably deep—level. I believe there may be a new *kind* of mathematical theory out there in the intellectual darkness; I believe that biology is the key to finding it. I fervently hope that if we *can* find it, then it will greatly illuminate some of the big puzzles of biology—development, reproduction, ecosystem dynamics, evolution. . . . But whether it does or not, it will certainly give the mathematicians a lot to chew on, and it will have spin-offs in the physical sciences. I even have a name for it: "morphomatics." Names, however, are cheap; I have no clear idea of how to go about setting morphomatics up, let alone of understanding it. Isaac Newton's wonderful achievement was to invent a new kind of mathematics—calculus—*and* to invent enough machinery to be able to use it to solve real problems. I have no claims to being a Newton—I can't even invent the mathematics, let alone develop it, and absolutely not apply it.

But I can see that it might be there. Let me start with some ingredients that *are* present now. I'm not so much interested in the details as in the philosophical stance and the conceptual viewpoint. Let me explain, and you'll see what I mean by that.

One of the most significant developments is that mathematics in general has become more geometric. Not the rigid geometry of Euclid: the visual geometry of the mind's eye. One important consequence is that qualitative reasoning has been put on a formal basis and turned into a precise tool. The physicist Ernest Rutherford once remarked that "qualitative is just poor quantitative," and this derogatory remark seems to have stuck, which is a terrible shame. In its proper context, that of numerical measurements in experiments, Rutherford's distaste for the qualitative is reasonable enough; it is much more impressive to predict, say, that 6.73 cm (about 2½ inches) of rain will fall tomorrow, than just to predict "heavy rainfall." However, there are many circumstances in which qualitative information is what really counts, and quantitative measurements are a rather poor route toward finding that information. For example, the most important question about a bridge is, "will it fall down?" and calculating its precise numerical breaking strain on a supercomputer is just an exceedingly complicated way to get a yes/no answer—and perhaps an unnecessarily complicated one. . . . If only we knew a better way, which in this case, we don't.

Much more significantly, there are occasions when a system is so complicated that quantitative information is largely useless. Because of major advances in instrumentation, it is commonplace nowadays for scientists to collect huge quantities of data—for example, images of brain function while a person is smelling a rose or looking at a picture. Having done so, the question arises of working out what those data are telling us—and here the raw numbers are pretty useless. Somehow, the significant patterns must be extracted, and that, in effect, is an attempt to reduce the quantitative data to its important qualitative features. So the word *qualitative* is used in at least two ways. To Rutherford, it meant "vague generalities." To today's mathematicians, however, it means "features that are conceptually deeper than mere numbers."

One of the simplest, yet most informative, ideas that this kind of thinking has given us is the concept of a *phase space*—the space of all possible things that *might* happen to a system, rather than just the stuff that it happens to be doing right now. What this does is to force us to rephrase questions about individual behaviors and systems, and to embed them in a richer context. Instead of looking at one water wave and wondering why it does what it does, we look at an entire space of possible shapes and movements for water, seek relationships among them, and work out how simple natural rules pick out the behavior that actually occurs. There is a growing tendency for biologists to talk of "morphospaces" inhabited by possible designs of organisms, and "DNA-space" in which all possible DNA sequences are arranged side by side to show what their similarities are. I applaud the use of such mathematical imagery, and I hope the trend accelerates.

There are many other deep, simple principles: continuity, connectivity, feedback, information, order, disorder, bifurcation, learning, autonomy, emergence. . . . A deep principle of special relevance to pattern formation is symmetry. Our universe has fundamental symmetries; one consequence is that we live in a world of patterns. During the 1900s, mathematicians have developed a kind of symmetry calculus that lets them reason about patterns on a general level, without worrying much about their precise internal details. This way of thinking makes it rather obvious, for instance, that patterns in the BZ reaction and patterns in slime mold *should* bear a strong resemblance to each other.[3]

Big, abstract, deep mathematical principles of this kind have got to be the things that drive morphomatics. Nature may build patterns by some intricate reductionist chain of activity, but it chooses which ones to retain or to modify in terms of their useful features. Pattern selection happens on the level of *the pattern as a whole.* When a cat is stalking a bird, it doesn't know what genes the bird has, but it can easily tell whether the bird is slow to fly away when danger threatens. The genes may be causing that slowness, but it is the slowness itself that is apparent to the cat.

Right now, we are aware of a few of these deep principles. We need more. We also need a better understanding of how to use those principles. We also need to extend the range of systems that our mathematics can handle. But—as I've tried to convince you—we're getting there. D'Arcy Thompson would have loved to be alive today, when his ideas are starting to bear fruit. He would have loved to learn about complexity, chaos, fractals, genetic algorithms, neural nets, and cellular automata. And I suspect he would be even happier to have been alive in a few centuries' time, when finally the human race manages to develop a unified theory of the deep mathematical laws behind growth and form.

Life's *other* secret.

Coming soon to a planet near you.

Notes

The main purpose of the notes is to provide precise references for cited work, but there are also extra arguments, counterarguments, anecdotes, comments . . . whatever.

1: What Is Life?

1. For details about the discoveries suggesting the former existence of Martian life, see the special report "Life on Mars," *New Scientist* (August 17, 1996), 4–11.

2. The word *law,* though standard, reflects the outmoded view that the mathematical patterns that humans find in nature are *exact.* All currently known laws of nature are approximations. For all we know, nature may not obey *exact* laws at all. On the other hand, there may be a single theory of everything that represents the true laws of nature. The jury is still out.

3. The equation in question is called the Schrödinger Equation, after its discoverer, Erwin Schrödinger. For an accessible discussion of quantum mechanics, see John Gribbin, *In Search of Schrödinger's Cat,* Black Swan, London (1991).

4. D'Arcy Wentworth Thompson, *On Growth and Form* (2 volumes, second edition), Cambridge University Press, Cambridge, England (1942). An abridged edition: D'Arcy Wentworth Thompson, *On Growth and Form* (edited by J. T. Bonner), Cambridge University Press, Cambridge, England (1961).

5. James Watson, *The Double Helix,* Signet, New York (1968).

6. Adrian L. R. Thomas and Andrew Balmford, "How natural selection shapes birds' tails," *The American Naturalist* **146** (1995), 848–868.

2: Before Life Began

1. Johannes Kepler, *The Six-Cornered Snowflake (De Nive Sexangula),* edited and translated by Colin Hardie, Oxford University Press, Oxford, England (1976).

2. Roger Davey and David Stanley, "All about ice," *New Scientist* (December 18, 1993), 33–37.

3. Peter J. Marchand, "Waves in the forest," *Natural History 2* (1995), 26–33.

4. Andrew Goudie, *The Nature of the Environment* (second edition), Blackwell, Oxford, England (1989).

5. Ian Stewart and Martin Golubitsky, *Fearful Symmetry,* Blackwell, Oxford, England (1992); Penguin, Harmondsworth, England (1993).

6. Cohen and Winfree's simplification of Zhabotinskii's recipe works reliably even in front of students, and uses only four cheap components having a shelf life of years, especially if they are refrigerated. It produces bromine, but not in dangerous amounts in an airy room. *Caution:* The mixture is moderately poisonous.

   Make up four components:

   a. 25 grams of sodium bromate, 335 milliliters (ml) water to dissolve the bromate, then 10 ml concentrated sulphuric acid.

   b. 10 grams of sodium bromide, and water to 100 ml.

   c. 10 grams of malonic acid, and water to 100 ml.

   d. 1, 10 phenanthroline ferrous complex (Fisons, Loughborough, United Kingdom).

   Put 6 ml of solution A in a glass beaker, then add 0.5 ml of solution B, then quickly mix in 1 ml of solution C. Leave the brown mixture by an open window to lose bromine, until it is the color of pale straw or is colorless (2–3 minutes if agitated in a flat dish).

Add 1 ml of the redox indicator D, mix thoroughly, and pour into a 9-centimeter glass or plastic petri dish on a white (preferably illuminated) background.

It will turn patchy blue, then clear to a brownish-red. Foci of blue will appear (you may have to wait up to 5 minutes) and grow into a series of concentric rings, expanding slowly. If the dish is shaken to restore homogeneity, the patterns reappear. Otherwise, do not jar or vibrate the dish. The effect lasts for 20 to 25 minutes. The experiment works very well on an overhead projector, for visibility in a classroom, but the rings may be fuzzy if the cooling fan is unbalanced.

7. M. Golubitsky, E. Knobloch, and I. Stewart, "Spirals in reaction-diffusion systems," preprint, Mathematics Department, University of Houston (1997).

8. Ian Stewart and Martin Golubitsky, *Fearful Symmetry,* Blackwell, Oxford, England (1992); Penguin, Harmondsworth, England (1993).

9. The point of sperm entry into the egg plays a role in determining *where* the buckling occurs.

3: The Frozen Accident

1. Stuart A. Kauffman, *The Origins of Order,* Oxford University Press, Oxford, England (1993).

2. A. G. Cairns-Smith, *Seven Clues to the Origin of Life,* Cambridge University Press, Cambridge, England (1985).

3. James Watson, *The Double Helix,* Signet, New York (1968).

4. Richard R. Sinden, *DNA Structure and Function,* Academic Press, San Diego (1994).

5. Exceptions include fungi and many bacteria and archaea. In addition, mitochondria and plastids have their own genetic code, different from that of the surrounding cell—and from each other.

6. José Hornos and Yvonne Hornos, *Physics Review Letters* **71** (1993), 4401–4404.

7. Ian Stewart, *From Here to Infinity,* Oxford University Press, Oxford, England (1996).

8. Gina Kolata, "Solving knotty problems in math and biology," *Science* **231** (1986), 1506–1508; De Witt Sumners, "Lifting the curtain: Using topology to probe the hidden action of enzymes," *Notices of the American Mathematical Society* **42** (1995), 528–537.

9. John L. Casti and Anders Karlqvist (editors), *Boundaries and Barriers,* Addison-Wesley, Reading, Mass. (1996); Martin Karplus and J. Andrew McCammon, "The dynamics of proteins," *Scientific American* (April 1986), 30–37.

10. P. Jonathan, G. Butler, and Aaron Klug, "The assembly of a virus," *Scientific American* (November 1978), 52–59.

11. James M. Hogle, Marie Chow, and David J. Filman, "The structure of poliovirus," *Scientific American* (March 1987), 28–35.

12. H. S. M. Coxeter, "Virus macromolecules and geodesic domes," *A Spectrum of Mathematics: Essays presented to H. G. Forder* (edited by John Butcher), Oxford University Press, Oxford, England (1967); Ian Stewart, *Game, Set and Math,* Penguin Books, Harmondsworth, England (1989).

4: The Oxygen Menace

1. For an extensive discussion of this point of view on free will, see Ian Stewart and Jack Cohen, *Figments of Reality,* Cambridge University Press, Cambridge, England (1997), Chapter 9.

2. Christian de Duve, "The birth of complex cells," *Scientific American* (April 1996), 38–45.

3. David M. Glover, Cayetano Gonzalez, and Jordan W. Raff, "The centrosome," *Scientific American* (June 1993), 32–38; Eric Bailly and Michael Bornens, "Centrosome and cell division," *Nature* **355** (1992), 300–301.

4. B. C. Goodwin, "Developing organisms as self-organizing fields," in *Mathematical Essays on Growth and the Emergence of Form* (edited by Peter L. Antonelli), University of Alberta Press, Alberta, Canada (1982), 185–200; B. C. Goodwin and Norbert H. J. Lacroix, "A further study of the holoblastic cleavage field," *Journal of Theoretical Biology* **109** (1984), 41–58; Brian C.

Goodwin and Stuart A. Kauffman, "Spatial harmonics and pattern specification in early *Drosophila* development: Part I. Bifurcation sequences and gene expression," *Journal of Theoretical Biology* **144** (1990), 303–319.

5. Maarten C. Boerlijst, *Selfstructuring: A Substrate for Evolution*, Ph.D. thesis, University of Utrecht (1994); Thomas Höfer, *Modelling* Dictyostelium *Aggregation*, Ph.D. thesis, Balliol College, Oxford University (1996).

6. In the United States, this is called a "rotary" or a "traffic circle."

7. M. Golubitsky, E. Knobloch, and I. Stewart, "Spirals in reaction-diffusion systems," preprint, Mathematics Department, University of Houston (1997).

5: Artificial Life

1. Charles Darwin, *The Origin of Species*, Penguin Books, Harmondsworth, England (1985).

2. D. Lack, *Darwin's Finches*, Cambridge University Press, Cambridge, England (1947).

3. Mendel's story is told, in some detail, in Karl Sigmund, *Games of Life*, Oxford University Press, Oxford, England (1993).

4. The DNA point of view is brilliantly advocated by Richard Dawkins in his books *The Blind Watchmaker*, Longman, London (1986); *The Selfish Gene*, Oxford University Press, Oxford, England (1989); *The Extended Phenotype*, Oxford University Press, Oxford, England (1982); *River out of Eden*, Weidenfeld and Nicolson, London, England (1995); and *Climbing Mount Improbable*, Viking, London (1996).

5. Mark Ridley, *Evolution*, Blackwell Science, Oxford, England (1996).

6. V. Sarich and A. C. Wilson, "Immunological timescale for human evolution," *Science* **158** (1967), 1200–1203.

7. C. G. Sibley and J. E. Ahlquist, "DNA hybridization evidence of hominoid phylogeny: Results from an expanded data set," *Journal of Molecular Evolution* **26** (1987), 99–121.

8. See, for example, John Maynard Smith, *Evolution and the Theory of Games*, Cambridge University Press, Cambridge, England (1978).

9. Sewall Wright, *Evolution: Selected Papers* (edited by W. B. Provine), University of Chicago Press, Chicago (1986).

10. Jack Cohen and Ian Stewart, *The Collapse of Chaos*, Viking, New York (1994).

11. E. C. Zeeman, "Evolution and elementary catastrophes," preprint, Mathematics Institute, University of Warwick, England (1988).

12. N. Eldredge and S. J. Gould, "Punctuated equilibria: An alternative to phyletic gradualism," in T. J. M. Schopf (editor), *Models in Palæobiology*, Freeman, Cooper & Co., San Francisco (1972).

13. René Thom, *Structural Stability and Morphogenesis* (translated by David H. Fowler), Benjamin-Addison-Wesley, New York (1975); Tim Poston and Ian Stewart, *Catastrophe Theory and Its Applications*, Pitman, London (1978); E. C. Zeeman, *Catastrophe Theory: Selected Papers 1972–77*, Addison-Wesley, Reading, Mass. (1977).

14. Roger Lewin, *Complexity*, Macmillan, New York (1992); Mitchell Waldrop, *Complexity*, Simon and Schuster, New York (1992); Stuart A. Kauffman, *The Origins of Order*, Oxford University Press, Oxford, England (1993); Klaus Mainzer, *Thinking in Complexity*, Springer-Verlag, Berlin (1994); Stuart A. Kauffman, *At Home in the Universe*, Viking, New York (1995).

15. "Life and death in a digital world," *New Scientist* (February 22, 1992), p. 36; Ed Regis, *Great Mambo Chicken and the Transhuman Condition*, Addison-Wesley, Reading, Mass. (1990).

16. Paul Guinnessy, "Life crawls out of the digital soup," *New Scientist* (April 13, 1996), p. 16.

6: Flowers for Fibonacci

1. Przemyslaw Prusinkiewicz and Aristid Lindenmayer, *The Algorithmic Beauty of Plants*, Springer-Verlag, New York (1990).

2. The golden number is often taken to be $(\sqrt{5} + 1)/2$, with a plus sign, which equals 1.618034. This is the same as both $1+\varphi$ and $1/\varphi$, and it's a matter of convention which one you call the golden number. If you look at the ratio the other way up, say 55/34 = 1.6176, then in the limit, you get 0.618034 instead.

3. H. Vogel, "A better way to contruct the sunflower head," *Mathematical Biosciences* **44** (1979), 145–174.

4. To see why, suppose the divergence angle were 180°, which divides 360° exactly. Then successive primordia would be arranged along two radial lines, diametrically opposite each other. If you used a divergence angle of 90°, you'd get four radial lines. In fact, if you use a divergence angle that is a rational multiple of 360° — an angle $360p/q$ for whole numbers p and q—then you get q radial lines. That means there will be big gaps in the arrangement, between those radial lines. So the seeds don't pack efficiently. To get efficient filling of the space you need a divergence angle that is an irrational multiple of 360°—a multiple by a number that is not an exact fraction. In fact, you ought to use the most irrational number possible—and number theorists have long known that the most irrational number is the golden number. That may sound like a strange description: Numbers are either irrational or not—but some are more irrational than others. Recall that the golden number φ is the limit of the sequence 2/3, 3/5, 5/8, 8/13, 13/21, 21/34, 34/55, and so on. Those are rational approximations to φ: rational numbers that get closer and closer to it but never equal it exactly. Now, we can measure how irrational φ is by seeing how quickly the errors—the differences between these fractions and φ—shrink toward zero. What the number theorists proved is that for φ, the differences shrink more slowly than they do for any other irrational number (see A. Ya. Khinchin, *Continued Fractions*, Phoenix, University of Chicago Press, Chicago, 1964, p. 36). The golden number is badly approximable by rationals, and if you quantify the degree of approximation in a reasonable way, it's the worst of them all. So the golden number is *special*—it can be singled out from every other number there is by a simple mathematical property. In essence, that's the argument that Vogel came up with.

5. Stéphane Douady and Yves Couder, "La physique des spirales végétales," *La Recherche* **24** (January 1993), 26–35; "Phyllotaxis as a self-organized growth process," in *Growth Patterns in Physical Sciences and Biology* (edited by J. M. García-Ruiz et al.), Plenum, New York (1993), 341–352; "Phyllotaxis as a self-organized growth process," *Physical Review Letters* **68** (1992), 2098–2101.

6. M. Kunz, "Some analytical results about two physical models of phyllotaxis," *Communications in Mathematical Physics* **169** (1995), 261–295.

7. Przemyslaw Prusinkiewicz and Aristid Lindenmayer, *The Algorithmic Beauty of Plants,* Springer-Verlag, New York (1990).

7: Morphogens and Mona Lisas

1. For a quick tour of the remarkable variety of seashell patterns, see R. Tucker Abbott, *Seashells of the World,* Golden Press, New York (1985).

2. Hans Meinhardt, *The Algorithmic Beauty of Sea Shells,* Springer-Verlag, Berlin (1995).

3. D. M. Raup, "Computer as aid in describing form of gastropod shells," *Science* **138** (1962), 150–152; C. Illert, "Formulation and solution of the classical seashell problem," *Il Nuovo Cimento* **11** D (1989), 761–780.

4. A. M. Turing, "The chemical basis of morphogenesis," *Philosophical Transactions of the Royal Society of London,* series B **237** (1952), 37–72.

5. J. D. Murray, *Mathematical Biology,* Springer-Verlag, New York (1989), p. 444.

6. Lewis Wolpert, "The shape of things to come," *New Scientist* (June 27, 1992), 38–42.

7. J. D. Murray, *Mathematical Biology,* Springer-Verlag, New York (1989), p. 448.

8. For a popular exposition, see Brian Goodwin, *How the Leopard Changed Its Spots,* Weidenfeld and Nicolson, London (1994). The technical literature includes B. C. Goodwin and L. E. H. Trainor,

"Tip and whorl morphogenesis in *Acetabularia* by calcium-regulated strain fields," *Journal of Theoretical Biology* **117** (1985), 79–106; C. Brière and B. C. Goodwin,"Effects of calcium input/output on the stability of a system for calcium-regulated viscoelastic strain fields," *Journal of Mathematical Biology* **28** (1990), 585–593; M. A. J. Chaplain and B. D. Sleeman, "An application of membrane theory to tip morphogenesis in *Acetabularia*," *Journal of Theoretical Biology* **146** (1990), 177–200; Brian C. Goodwin and Stuart A. Kauffman, "Spatial harmonics and pattern specification in early *Drosophila* development: Part I. Bifurcation sequences and gene expression," *Journal of Theoretical Biology* **144** (1990), 303–319; Axel Hunding, Stuart A. Kauffman, and Brian C. Goodwin, *Drosophila* segmentation: Supercomputer simulation of prepattern hierarchy," *Journal of Theoretical Biology* **145** (1990), 369–384; B. C. Goodwin and C. Brière, "A mathematical model of cytoskeletal dynamics and morphogenesis in *Acetabularia*," in *The Cytoskeleton of the Algae* (edited by Diedrik Menzel), CRC Press, Boca Raton, Fla. (1992), 219–237.

9. Ironically, chemists have now managed to create stationary Turing patterns, in gels. See V. Castets, E. Dulos, J. Boisonnade, and P. de Kepper, *Physics Review Letters* **64** (1990), 2953–2956; Q. Ouyang and Harry L. Swinney, "Transition from a uniform state to hexagonal and striped Turing patterns," *Nature* **352** (1991), 610–612.

10. Shigeru Kondo and Rihito Asai, "A reaction-diffusion wave on the skin of the marine angelfish *Pomacanthus*," *Nature* **376** (1995), 765–768; Hans Meinhardt, "Dynamics of stripe formation," *Nature* **376** (1995), 722–723.

11. Lewis Wolpert, *The Triumph of the Embryo,* Oxford University Press, Oxford, England (1991).

12. It is known that relatively small changes in the *time* at which the protostripes become imprinted on the zebra embryo produce big changes in the pattern on the adult; in fact, different species of zebra, with very different adult patterns, trace back to the same embryonic structure—but separated by a few hours of development. See J. B. L. Bard, "A unity underlying the different zebra

striping patterns," *Journal of the Zoological Society of London* **183** (1977), 527–539; J. D. Murray, *Mathematical Biology,* Springer-Verlag, New York (1989), p. 442.

13. Pankaj Agarwal, "The cell programming language," *Artificial Life* **2** (1995), 37–77.

8: The Peacock's Tale

1. Klaus Mainzer, *Thinking in Complexity,* Springer-Verlag, Berlin (1994).

2. John H. Hopfield and David W. Tank, "Computing with neural circuits: A model," *Science* **233** (1986), 625–633.

3. Sahley and Gelperin's work is described by Georgina Ferry, "Networks on the brain," *New Scientist* (July 16, 1987), 54–58; See also Jules Davidoff and David Concar, "Brain cells made for seeing," *New Scientist* (April 10, 1993), 32–36; Tobias Bonhoeffer and Amiram Grinvald, "Iso-orientation domains in cat visual cortex are arranged in pinwheel-like patterns," *Nature* **353** (1991), 429–431; Steven P. Dear, James A. Simmons, and Jonathan Fritz, "A possible neuronal basis for representation of acoustic scenes in auditory cortex of the big brown bat," *Nature* **364** (1993), 620–623.

4. Simon Clippingdale and Roland Wilson, "Self-similar neural networks based on a Kohonen learning rule," preprint, Computer Science Department, University of Warwick, England (1994); Peter Mason, Ph.D. thesis, Mathematics Institute, University of Warwick, England (1996).

5. J. D. Cowan, "Brain mechanisms underlying visual hallucinations," preprint, Mathematics Department, University of Chicago; "Spontaneous symmetry breaking in large scale nervous activity," *International Journal of Quantum Chemistry* **22** (1982), 1059–1082.

6. David Concar, "Sex and the symmetrical body," *New Scientist* (April 22, 1995), 40–44; Mark Kirkpatrick and Gil G. Rosenthal, "Symmetry without fear," *Nature* **372** (1994), 134–135; Matt Ridley, "Swallows and scorpionflies find symmetry is beautiful,"

Science **257** (1992), 327–328; R. Thornhill, S. W. Gangstead, and R. Comer, "Human female orgasm and mate fluctuating asymmetry," *Animal Behaviour* **50** (1995), 1601–1615; David Palliser, "Symmetry in the human body is sexier and healthier as well as aesthetically pleasing, says scientist," *The Guardian* (August 10, 1996), p. 4.

7. Magnus Enquist and Anthony Arak, "Symmetry, beauty and evolution," *Nature* **372** (1994), 169–172; Magnus Enquist and Anthony Arak, "The illusion of symmetry?" *The Journal of NIH Research* **7** (July 1995), 54–55.

8. Rufus A. Johnstone, "Female preference for symmetrical males as a by-product of selection for mate recognition," *Nature* **372** (1994), 172–175. See also Martin Giurfa, Birgit Eichmann, and Randolf Menzel, "Symmetry perception in an insect," *Nature* **382** (1996), 458–461.

9: Walk on the Wild Side

1. Joh Henschel, "Spider revolutions," *Natural History* **3** (1995), 36–39.

2. David P. Maitland, "Locomotion by jumping in the Mediterranean fruit-fly larva *Ceratitis capitata*," *Nature* **355** (1992), 159–160.

3. P. P. Gambaryan, *How Mammals Run,* Wiley, New York (1974).

4. M. Hildebrand, "Symmetrical gaits of horses," *Science* **150** (1965), 701–708.

5. G. Schöner, W. Y. Jiang, and J. A. S. Kelso, "A synergetic theory of quadrupedal gaits and gait transitions," *Journal of Theoretical Biology* **142** (1990), 359–391; J. J. Collins and I. N. Stewart, "Coupled nonlinear oscillators and the symmetries of animal gaits," *Journal of Nonlinear Science* **3** (1993), 349–392.

6. Some horses, however, specially bred in the United States, pace but never trot.

7. J. J. Collins and I. N. Stewart, "Hexapodal gaits and coupled nonlinear oscillator models," *Biological Cybernetics* **68** (1993), 287–298.

8. M. Golubitsky, I. Stewart, J. J. Collins, and L. Buono, "A modular network for legged locomotion," preprint, Mathematics Department, University of Houston (1997).

9. Allyn Jackson, "Lamprey lingo," *Notices of the American Mathematical Society* **38** (1991), 1236–1239; Sten Grillner, "Neural networks for vertebrate locomotion," *Scientific American* (January 1996), 48–53.

10. A. H. Cohen, P. J. Holmes, and R. H. Rand, "The nature of the coupling between segmental oscillators of the lamprey spinal generator for locomotion: a mathematical model," *Journal of Mathematical Biology* **13** (1982), 345–369.

10: An Exaltation of Boids

1. Fritz Vollrath, "Spider webs and silks," *Scientific American* (March 1992), 52–58; Kate Douglas, "Arachnophilia," *New Scientist* (August 10, 1996), 24–28.

2. John H. Holland, "Genetic algorithms," *Scientific American* (July 1992), 44–50.

3. Douglas R. Hofstadter, *Gödel, Escher, Bach: An Eternal Golden Braid,* Penguin Books, Harmondsworth, England (1980).

4. R. V. Solé, O. Miramontes, and B. C. Goodwin, "Emergent behaviour in insect societies: Global oscillations, chaos and computation," in *Interdisciplinary Approaches to Nonlinear Complex Systems* (edited by H. Haken and A. Mikhailov), Springer series in Synergetics **62**, Springer-Verlag, Berlin (1993), 77–88; Ricard V. Solé, Octavio Miramontes, and Brian C. Goodwin, "Oscillations and chaos in ant societies," *Journal of Theoretical Biology* **161** (1993), 434–357; Octavio Miramontes, Ricard V. Solé, and Brian C. Goodwin, "Collective behaviour of random-activated mobile cellular automata," *Physica* D **63** (1993), 145–160.

5. C. W. Reynolds, "Flocks, herds, and schools: A distributed behavioral model," *Computer Graphics* **21** (part 4) (1987), 25–34. "Boid" is Bronx for "bird": as two anonymous New York poems put it:

> *Toity poiple boids*
> *Sitt'n on der coib*
> *A' choipin' an' a' boipin'*
> *An' eat'n doity woims.*

And:

> *Der spring is sprung*
> *Der grass is riz*
> *I wonder where dem boidies is?*
> *Der little boids is on der wing.*
> *Ain't dat absoid?*
> *Der little wings is on der boid.*

6. Jessica K. Hodgins and David C. Brogan, "Robot herds: group behaviors for systems with significant dynamics," in *Artificial Life IV*, MIT Press, Cambridge, Mass. (1994), 319–324.

7. Sean Blair, "The secret of crowds," *Focus* (June 1996), 26–29.

8. The classic trilogy is Isaac Asimov, *Foundation, Foundation and Empire, Second Foundation*, all of which originally appeared in the magazine *Astounding Science Fiction* (in, respectively, 1942–1944, 1945, and 1948–1950). They appeared in book form in 1951, 1952, and 1953 and are still in print from various publishers. Then there are much more modern sequels: *Foundation's Edge* (Doubleday, 1981), *Foundation and Earth* (Doubleday, 1985), *Prelude to Foundation* (Doubleday, 1987), *Forward the Foundation* (Doubleday, 1993).

9. For a flavor of *Red Dwarf*, see Grant Naylor, *Primordial Soup: Red Dwarf, the Least Worst Scripts*, Penguin Books, Harmondsworth, England (1993).

10. In 1996, Newton Da Costa and C. Doria announced a proof that the relevant mathematical problem, known as "P ≠ NP," is formally undecidable. The proof has not appeared at the time of this writing.

11: Reef Wars

1. J. von Neumann, *Theory of Self-Reproducing Automata*, University of Illinois Press, Urbana, 1966.

2. I can't help wondering whether von Neumann subconsciously got his idea from the work of Kurt Gödel and Alan Turing, who proved that certain mathematical questions exist, to which neither "yes" nor "no" is a provable answer. Certainly, he knew that their work rests on exactly the same dual interpretation of information: first as a sequence of instructions that has meaning within some chosen system, then as a string of symbols with no meaning whatsoever. This distinction, and the link between Gödel and DNA, is one of many brilliant passages in Douglas R. Hofstadter, *Gödel, Escher, Bach: An Eternal Golden Braid*, Penguin Books, Harmondsworth, England (1980).

3. S. Wolfram, *Theory and Applications of Cellular Automata*, World Scientific, Singapore (1986); Melanie Mitchell, "Computation in cellular automata: A selected review," preprint 96-09-074, Santa Fe Institute (1996). See also Christopher G. Langton, "Studying artificial life with cellular automata," *Physica* D **22** (1986), 120–149.

4. J. M. McGlade and A. R. G. Price, "Multi-disciplinary modelling: an overview and practical implications for the governance of the gulf region," *Marine Pollution Bulletin* **27** (1993), 361–377.

5. Ronald L. Graham, Donald E. Knuth, and Oren Patashnik, *Concrete Mathematics*, Addison-Wesley, Reading, Mass. (1994), pp. 290–301, especially equation (6.123), p. 299.

6. V. Courtillot and Y. Gaudemer, "Effects of mass extinctions on biodiversity," *Nature* **381** (1996), 146–148. For a description of one of the most important mass extinctions, see Jeffrey S. Levinton, "The Big Bang of animal evolution," *Scientific American* (November 1992), 52–59.

7. Ian Stewart and Jack Cohen, *Figments of Reality*, Cambridge University Press, Cambridge, England (1997).

8. Ian Stewart, *Does God Play Dice?* (second edition), Penguin, Harmondsworth, England (1997).

12: In Search of Secrets

1. Jack Cohen and Ian Stewart, *The Collapse of Chaos*, Viking, New York (1994).

2. Jack Cohen and Ian Stewart, *The Collapse of Chaos*, Viking, New York (1994); Ian Stewart and Jack Cohen, *Figments of Reality*, Cambridge University Press, Cambridge, England (1997).

3. Ian Stewart and Martin Golubitsky, *Fearful Symmetry*, Blackwell, Oxford, England (1992); Penguin, Harmondsworth, England (1993).

Further Reading

Each entry is annotated with a minireview. (Other people reviewed my own books.)

R. Tucker Abbott, *Seashells of the World,* Golden Press, New York (1985). [I bought this little paperback in the Houston Science Museum: It's delightful, with color illustrations of hundreds of seashells.]

A. G. Cairns-Smith, *Seven Clues to the Origin of Life,* Cambridge University Press, Cambridge, England (1985). [An accessible account of clay as the origin of life.]

Jack Cohen, *The Privileged Ape: Cultural Capital in the Making of Man,* Parthenon, Carnforth, England (1989). [An account of human evolution, emphasizing the role of privilege; sometimes turgid.]

Jack Cohen, *Reproduction,* Butterworths, London (1977). [A classic student text collecting all aspects of biological reproduction between two covers; highly illustrated.]

Jack Cohen and Brendan Massey, *Animal Reproduction: Parents Making Parents,* Edward Arnold, London (1984). [A simplified version of Cohen's *Reproduction.*]

Jack Cohen and Ian Stewart, *The Collapse of Chaos,* Viking, New York (1994). [How do complex structures exist in a chaotic world?]

Charles Darwin, *The Origin of Species,* Penguin Books, Harmondsworth, England (1859/1985). [Still very worth reading.]

Richard Dawkins, *The Blind Watchmaker,* Longman, London (1986). [A beautifully written account of modern evolutionary theory—marred only by its basic premise that genes map to characters.]

Richard Dawkins, *Climbing Mount Improbable,* Viking, London (1996). [This time, the evolutionary image is mountaineering: You can either climb the steep face at the front, or walk up the long, shallow slope at the back.]

Richard Dawkins, *The Extended Phenotype,* Oxford University Press, Oxford, England (1982). [Sparkling and extremely clever backtrack on and amplification of *The Selfish Gene.*]

Richard Dawkins, *River out of Eden,* Weidenfeld and Nicolson, London (1995). [Short and snappy summary based around the image of the river of evolution.]

Richard Dawkins, *The Selfish Gene* (second edition), Oxford University Press, Oxford, England (1989). [Elegant account of neo-Darwinism; argues the view that DNA rules.]

Daniel C. Dennett, *Consciousness Explained,* Little, Brown & Co., Boston (1991). [Utterly brilliant, nonreductionist. Critics who call it *Consciousness Explained Away* haven't understood the nature of the problem.]

Daniel C. Dennett, *Kinds of Minds,* Nicolson, London (1996). [A gentler discussion of the same ideas, with an added dash of human culture.]

Jared Diamond, *The Rise and Fall of the Third Chimpanzee,* Vintage, London (1991). [Why are humans so different from chimps when they share 98% of the same DNA? Argues, compellingly and in fascinating detail, that most differences have precursors elsewhere in the animal kingdom.]

Robert Foley, *Humans before Humanity,* Blackwell, Oxford (1995). [How a gang of apes were challenged by progressive deforestation and became naked, sweaty humans.]

Brian Goodwin, *How the Leopard Changed Its Spots,* Weidenfeld and Nicolson, London (1994). [Why development and evolution are dynamic phenomena, not just rampant genes. Required reading from one of the world's most innovative biologists.]

Stephen Jay Gould, *Ever Since Darwin*, Penguin Books, Harmondsworth, England (1980); *The Mismeasure of Man*, Penguin (1981); *The Panda's Thumb*, Penguin (1983); *Hen's Teeth and Horse's Toes*, Penguin (1984); *Time's Arrow, Time's Cycle*, Penguin (1988); *An Urchin in the Storm*, Penguin (1990); *Bully for Brontosaurus*, Penguin (1992). [Witty, pithy, memorable essays—you can learn better biology from them than from most textbooks.]

Stephen Jay Gould, *Wonderful Life*, Penguin Books, Harmondsworth, England (1991). [The famous account of the soft-bodied creatures of the Burgess shale and the celebration of contingency in evolution.]

Douglas R. Hofstadter, *Gödel, Escher, Bach: An Eternal Golden Braid*, Penguin Books, Harmondsworth, England (1980). [Classic mind-expanding cult book of the 1980s—funny, infuriating, and enormously illuminating; speculations on the emergent properties of ants' nests.]

Stuart A. Kauffman, *At Home in the Universe*, Viking, New York (1995). [A personal view of self-organization and the search for laws of complexity. Required reading from another of the world's most creative biologists.]

Stuart A. Kauffman, *The Origins of Order*, Oxford University Press, Oxford (1993). [A comprehensive and powerful discussion of self-organization as *the* motive force in evolution and development.]

Roger Lewin, *Complexity*, Macmillan, New York (1992). [A people-based description of the work of the Santa Fe Institute.]

Klaus Mainzer, *Thinking in Complexity*, Springer-Verlag, Berlin (1994). [Well-reasoned defense of the ideas behind complexity theory; strong on history and scientific content.]

Hans Meinhardt, *The Algorithmic Beauty of Sea Shells*, Springer-Verlag, Berlin (1995). [Profusely illustrated collection, in color, of seashell patterns and corresponding mathematical models produced by reaction-diffusion. If *this* doesn't convince you that D'Arcy Thompson had a point, nothing will. Don't even think about it: just go and buy it, now. Buy 10, and give them to your friends—especially the ones who take pride in their ignorance of mathematics.]

J. D. Murray, *Mathematical Biology*, Springer-Verlag, New York (1989). [A genuine textbook, this one, with all the formulas—but highly readable if you skip the technical bits; includes lots of pictures.]

Roger Penrose, *The Emperor's New Mind*, Oxford University Press, Oxford, England (1989). [Definitive source for "quantum uncertainty, therefore free will." Fun, unafraid of mathematical formulas, brilliant on almost everything. Pity that its central contention makes no sense at all.]

Roger Penrose, *Shadows of the Mind*, Oxford University Press (1994). [The sequel; challenging to nonspecialists, but very widely read, like its predecessor.]

Przemyslaw Prusinkiewicz and Aristid Lindenmayer, *The Algorithmic Beauty of Plants*, Springer-Verlag, New York (1990). [L-systems and their generalizations—the mathematics of the plant world; lots of marvelous illustrations, many in color.]

C. David Rollo, *Phenotypes*, Chapman and Hall, London (1994). [Stunning advocacy of the view that there is more to life than genes; worth buying just for the hopeful monster on the cover—a toad with eyes inside its mouth.]

Karl Sigmund, *Games of Life*, Oxford University Press, Oxford, England (1993). [The best popular book on game-theoretic models in ecology that currently exists—or is likely to.]

John Maynard Smith, *The Evolution of Sex*, Cambridge University Press, Cambridge, England (1982). [Games, genes, Red Queen, and evolutionarily stable strategies.]

John Maynard Smith, *Evolution and the Theory of Games*, Cambridge University Press, Cambridge, England (1978). [Much better than *The Evolution of Sex:* more contextual.]

Ian Stewart, *Does God Play Dice?* (second edition), Penguin Books, Harmondsworth, England (1997). [Superb general account of chaos, against a background of modern science.]

Ian Stewart, *Nature's Numbers*, Weidenfeld and Nicolson, London (1995). [A celebration of mathematical patterns in nature; short-listed for the 1995 Rhône-Poulenc Prize for Science Books.]

Ian Stewart and Jack Cohen, *Figments of Reality,* Cambridge University Press, Cambridge, England (1997). [The sequel to *The Collapse of Chaos:* How can animals like us, with minds like ours, come into being?]

Ian Stewart and Martin Golubitsky, *Fearful Symmetry,* Blackwell, Oxford, England (1992); Penguin, Harmondsworth, England (1993). [A whole new way of looking at pattern, complexity, and the generation of order in nature.]

D'Arcy Wentworth Thompson, *On Growth and Form* (2 volumes), Cambridge University Press, Cambridge, England (1942). [The great classic, full of thoughtful insights and thought-provoking examples.]

D'Arcy Wentworth Thompson, *On Growth and Form* (edited by J. T. Bonner), Cambridge University Press, Cambridge, England (1961). [Abridged edition, which conveys the same message in a more digestible form.]

Mitchell Waldro, *Complexity,* Simon and Schuster, New York (1992). [How emergence is becoming respectable: a detailed look at the Santa Fe Institute and the theories it is developing.]

James Watson, *The Double Helix,* Signet, New York (1968). [Insider's warts-and-all view of the discovery of the (first) secret of life.]

Scott Weidensaul, *Fossil Identifier,* Quintet, London (1992). [There were patterns in biology billions of years ago; this full-color book is packed with them.]

Art Wolfe and Barbara Sleeper, *Wild Cats of the World,* Crown, New York (1995). [A glorious reminder that nature still has the edge on mathematics when it comes to beautiful patterns and dynamic movement. (Look, I'm a cat person, OK?)]

Stephen Wolfram, *Theory and Applications of Cellular Automata,* World Scientific, Singapore (1986). [A collection of technical papers, still one of the best sources for the mathematics of cellular automata.]

Lewis Wolpert, *The Triumph of the Embryo,* Oxford University Press, Oxford, England (1991). [A leading biologist discusses biological development in terms of positional information.]

Semir Zeki, *A Vision of the Brain,* Blackwell Scientific, London (1993). [Excellent account of the eyes, the cortex, and perception of color and shapes.]

Credits

Figures

1: NASA.

2: Dawn Wright.

3, 21, 23, 24: Richard R. Sinden, *DNA Structure and Function,* Academic Press, San Diego, 1994.

4: H. G. Wells, *War of the Worlds,* as illustrated by Paul in a 1927 issue of *Amazing Stories.*

5, 6, 7, 8, 9, 10, 11, 34: D'Arcy W. Thompson, *On Growth and Form,* Cambridge University Press, 1992. Reprinted with the permission of Cambridge University Press.

12, 16, 19, 20, 22, 41, 42, 43, 44, 45, 52, 53, 63, 65, 68, 75, 77, 78, 79, 83: Drawn by Ian Stewart.

13, 64: Courtesy of *New Scientist.*

14: Andrew Goudie, *The Nature of the Environment,* Blackwell Publishers, 1989, pp. 131–132, Fig. 5.7.

15: G. Nicolis, *Introduction to Nonlinear Science,* Cambridge University Press, 1995. Reprinted with the permission of Cambridge University Press.

17: M. Shumway, *Anatomical Record, Vols. 78 and 83,* New York, 1942.

18: From J. M. T. Thompson and H.B. Stewart, *Nonlinear Dynamics and Chaos.* © John Wiley & Sons, Limited, 1986. Reproduced with permission.

25: Courtesy of Nicholas R. Cozzarelli.

26: Ronald Brown, "Out of line," *Royal Institution Proceedings, 64,* 207–243; Fig. 19 on page 223. Courtesy of Ronald Brown and T. Porter.

27, 28: J. L. Casti/A. Karlquist, *Boundaries and Barriers,* (Figs. 5 & 6 from page 27). © 1996 by John Casti. Reprinted by permission of Addison-Wesley Longman, Inc.

29: Courtesy of John T. Finch.

30, 31: Courtesy of Allen Beechel.

32: Ian Stewart, *Game, Set, & Math: Enigmas & Conundrums,* Blackwell Publishers, 1989.

33: Reprinted with permission of Dimitry Schildlovsky.

35: © David M. Phillips/Photo Researchers.

36, 37: C. Goodwin and Norbert H. J. Lacroix, "A further study of the holoblastic cleavage field," *Journal of Theoretical Biology, 109* (1984) 41–58. Courtesy of Academic Press.

38, 39, 40: Thomas Höfer, *Modelling* Dictyostelium *Aggregation,* Ph.D. thesis, Balliol College, Oxford University, 1996. Art courtesy of Thomas Höfer and Professor Peter Newell of Oxford University.

46: Courtesy of Jean Loup Charmet, Paris.

47, 48, 50, 51: Stéphane Douady and Yves Couder, "La physique des spirales végétales," *La Recherce, 24,* January 1993, 26–35.

49: Przemyslaw Prusinkiewicz and Aristid Lindenmayer, *The Algorithmic Beauty of Plants,* Springer-Verlag, New York, 1990.

54, 55: Przemyslaw Prusinkiewicz and Aristid Lindenmayer, *The Algorithmic Beauty of Plants,* Springer-Verlag, New York, 1990, p. 25, Fig. 1.24.

56: Brian C. Goodwin and Stuart A. Kauffman, "Spatial harmonics and pattern specification in early *Drosophila* development. Part I.

Bifurcation sequences and gene expression," *Journal of Theoretical Biology, 144* (1990) 303–319. Courtesy of Academic Press.

57: Q. Ouyang and Harry L. Swinney, "Transition from a uniform state to hexagonal and striped Turing patterns," *Nature, 352* (1991) 610–612. Reprinted with permission from *Nature.* © 1991 Macmillan Magazines Limited.

58: J. D. Murray, *Mathematical Biology,* Springer-Verlag, New York, 1989, Fig. 15.5, p. 444.

59, 81: Courtesy of Patricia J. Wynne.

60, 61, 62: Shigeru Kondo and Rihito Asai, "A reaction-diffusion wave on the skin of the marine angelfish *Pomacanthus,*" *Nature, 376* (1995) 765–768; p. 766, Fig. 1.

66: Wiley Photo Library.

67: Drawn by Ian Stewart based on Georgina Ferry, "Networks on the brain," *New Scientist* (16 July 1987), 54–58.

69: Courtesy of J. D. Cowan.

70: Sir James Gray, *Animal locomotion,* Weidenfield & Nicolson, London, 1968. See R. McNeil Alexander, "Swimming," in R. McNeil Alexander and G. Goldspink, *Mechanics and*

Energetics of Animal Locomotion, Chapman and Hall, London, 1977.

71, 73, 74: © Keter Publishing House Limited.

72: Courtesy of David P. Maitland.

76: J. J. Collins and I. N. Stewart, "Hexapodal gaits and coupled nonlinear oscillator models," *Biological Cybernetics, 68,* (1993) 287–292.

80: R. McNeil Alexander, "Terrestrial locomotion," in R. McNeil Alexander and G. Goldspink, *Mechanics and Energetics of Animal Locomotion*, Chapman and Hall, London, 1977.

82: Ricard V. Solé, Octavio Miramontes, and Brian C. Goodwin, "Oscillations and chaos in ant societies," *Journal of Theoretical Biology, 161* (1993) 434–357. Courtesy of Academic Press.

84: T. Toffoli and N. Margolus, *Cellular Automata Machines: A New Environment for Modeling,* MIT Press, 1987.

85: Reprinted with permission from *Nature.* V. Courtillot and Y. Gaudemer, "Effects of mass extinctions on biodiversity," *Nature 381* (1996), 146–148. Courtesy of Y. Gaudemer.

Plates

1: Alan Briere.

2: Courtesy of Jack Cohen.

3, 4: Courtesy of Thomas Höfer.

5: 1994/95, Christa Sommerer & Laurent Mignonneau, supported by ICC-NTT Japan and NCSA, Urbana IL, USA.

6: Courtesy of Stephanie Douady and Yves Couder.

7: Courtesy of David Fideler.

8, 9: Courtesy of Przemyslaw Prusinkiewicz.

10, 11, 16, 17: Tony Stone Images.

12, 13: Copyright © 1992 D. Fowler, H. Meinhardt, and P. Prusinkiewicz.

14, 15: Zigmund Leszczynski/ Animals Animals Enterprises, Inc.

18: David Overcash/Bruce Coleman Inc.

19: Courtesy of G. Keith Still.

20: Courtesy of J. M. McGlade.

Index